CONGENITAL DISORDERS - LABORATORY AND CLINICAL RESEARCH

CRANIOFACIAL DISORDERS

CAUSES, TYPES AND SURGICAL/TREATMENT OPTIONS

Congenital Disorders - Laboratory and Clinical Research

Additional books in this series can be found on Nova's website under the Series tab.

Additional e-books in this series can be found on Nova's website under the e-book tab.

CONGENITAL DISORDERS - LABORATORY AND CLINICAL RESEARCH

CRANIOFACIAL DISORDERS

CAUSES, TYPES AND SURGICAL/TREATMENT OPTIONS

SHARON B. TURNER
AND
JOSEPH E. MILLER
EDITORS

New York

For permission to use material from this book please contact us:
Telephone 631-231-7269; Fax 631-231-8175
Web Site: http://www.novapublishers.com

Library of Congress Cataloging-in-Publication Data

ISBN: 978-1-62618-923-2

LCCN: 2013938563

Published by Nova Science Publishers, Inc. † New York

Contents

Preface

Craniofacial abnormalities are congenital musculoskeletal disorders which primarily affect the cranium and facial bones. In this book, the authors present current research in the study of the causes, types and surgical/treatment options for craniofacial disorders. Topics include the etiopathogenesis, classification and treatment of temporomandibular joint disorders; paradigm shifts in pre-surgical orthopedics for infants with oral clefts; hearing loss in children with craniofacial anomalies; strabismus in craniosynostosis; and the orofacial features and treatment considerations of Cri du Chat Syndrome.

Chapter 1 – Temporomandibular joint (TMJ) disorders are a heterogeneous group of diseases that affects an increasing number of people, with 3-7% of the population seeking treatment for pain and dysfunction of the ATM. Apart from jaw pain, other common symptoms of TMJ disorders include limited or painful jaw movement, headache, neck pain or stiffness, clicking or grating within the joint, and, occasionally, an inability to open the mouth painlessly. The progressive joint degeneration leads finally to chronic pain and reduced quality of life. The etiology of TMJ disorders remains unclear, but it is likely multifactorial, with abnormal occlusion and emotional stress being considered the most serious etiological factors. Other risk factors includes age, gender, hyperlaxity, genetic factors, orthodontic treatment and parafunctional habits such as bruxism, teeth clenching, lip biting. The correct diagnosis of TMJ dysfunctions cannot be based on clinical examination only and several methods of imaging can be recommended: X ray examination, computer tomography, magnetic resonance imaging, radionuclide examination (scintigraphy), ultrasonography and arthroscopy. For most patients, the signs and symptoms of TMJ disorders improve over time with or without treatment. Therefore self-care and noninvasive treatments (e.g. acupuncture, hypnosis, temporary occlusal therapy, intra-articular corticosteroid or anesthetic injection, pharmacologic treatment, physical therapy modalities and psychological interventions such as cognitive behavior therapy, relaxation techniques or stress management) are good options and should be attempted before invasive or permanent therapies, such as orthodontics or surgery, are recommended. This chapter reviews the current knowledge in the etiopathogenesis, classification, diagnosis of TMJ disorders and discusses several surgical and non-surgical treatment protocols.

Chapter 2 – Pre-surgical orthopedic treatments have been employed in infants born with oral clefts for over half a century. There are a variety of technics described in the literature. The early treatment goals included feeding assistance as well as aligning alveolar segments. Advocates of the treatment claimed that long-term benefits included speech and orthodontic benefits. However, recent studies have shown that such long-term benefits could not be substantiated. The ability to mold neonatal cartilages has given rise to a new pre-surgical technic called Nasoalveolar Molding (NAM). The NAM technic added nasal cartilage molding as part of its treatment goal. This has allowed surgeons to consistently obtain improved esthetics in the nasolabial area. The popularity of NAM in recent years has shifted traditional treatment goals of the pre-surgical orthopedic treatments to focus more on enhancing nasolabial esthetics. In this chapter, the authors will discuss publications that deal with pre-surgical orthopedics for infants with oral cleft, particularly those dealing with NAM treatment.

Chapter 3 – It is well documented that children with craniofacial anomalies are at a greater risk of hearing impairment and middle ear pathology than children without craniofacial anomalies. Most research has focused on the hearing loss present in this population during the first few months of life. However, recent research has indicated that even if these infants pass newborn hearing screening at birth, on-going monitoring of hearing throughout childhood is required due to their increased risk of developing a postnatal hearing loss. A recent study completed by the authors revealed that children with craniofacial anomalies are 2.6 times more likely to develop a postnatal hearing loss than children without craniofacial anomalies. This chapter will begin with an overview of the type and degree of hearing impairment found in children with craniofacial anomalies and the implications of a hearing loss on the child's ability to develop speech and language. It will then explore methods to detect hearing loss from birth through childhood, with an emphasis on on-going monitoring. This chapter will conclude by addressing management options for hearing loss, which will help optimize the child's developmental outcomes.

Chapter 4 – Craniosynostosis is caused by abnormal premature fusion of any cranial suture. Because the orbits bridge the face to the cranium, ocular and adnexal abnormalities are common in many forms of craniosynostosis. Phenotypic classifications of craniosynostosis were initially developed, the simplest being non-syndromic craniosynostosis including scaphocephaly, trigonocephaly, synostotic plagiocephaly, versus syndromic craniosynostosis including the syndromes of Crouzon, Apert, Pfeiffer and Saethre-Chotzen. More recently, phenotypic classifications have been augmented by more mechanistic genetic studies of the fibroblast growth factor receptor genes, FGFR1, FGFR2, FGFR3 and TWIST genes, which have not corresponded well to phenotypic classifications.

Strabismus is observed in 39 to 91 % of patients with syndromic craniosynostosis. All types of strabismus may be seen. Mechanisms of strabismus in such cases are manifold, including heterotopy of muscle pulleys, absent muscles, and muscle force imbalance due to orbital asymmetry. Non-syndromic craniosynostosis are also associated with increased incidence of refractive errors and strabismus. Strabismus, ptosis, uncorrected refractive errors, and corneal exposure are all amblyogenic and can cause significant visual disability in children with craniosynostosis.

Diagnosis and treatment of strabismus and amblyopia in patients with craniosynostosis is important towards preservation of functional vision. Imaging provides objective parameters to determine appropriate strabismus surgery.

Improved neuro-surgical treatment in these patients has resulted in increased survival but there is evidence to show that the strabismus may change after such interventions. A variety of surgical maneuvers have been described to correct the strabismus. A multidisciplinary team comprising plastic surgeons, neurosurgeons, maxillofacial surgeons and ophthalmologists is valuable in preserving vision in these patients.

Chapter 5 – Cri du Chat syndrome (CdCS) is a genetic disease resulting from a partial or total deletion on the short arm of chromosome 5 (5p-). CdCS is one of the most frequent chromosomal syndromes in humans. The prevalence has been estimated at 1 in 37.000 live-born infants.

The most characteristic and clinical feature in newborns is a high-pitched monochromatic cat-like crying, that was attributed to anatomical alterations in the larynx and epiglottis. Additional clinical features are severe psychomotor and mental retardation, growth delay, and hypotonia. CdCS causes craniofacial abnormalities such as microcephaly, slight hypertelorism, epicanthal folds, downslanting palpebral fissures, strabismus, malformed ears, mandibular microretrognathism, anterior open-bite and high palate. Buccodental pathology and dental anomalies of patients with this syndrome reported are hypodontia, enamel hypoplasies, dental opacities, tooth agenesis, supernumerary tooth, root resortion, macrodontia, and transposed teeth.

Clinical dentistry management is linked to behavioral patterns. In general, CdC patients have euphoric, cheerful and sociable personalities. However, social isolation and reserved and emotionally detached character have been found in some cases of patients. Because of this and influenced for dental pathology and treatment, dental treatment is suggested to be ambulatory (without deep sedation) or, in contrast, with general anesthesia.

In: Craniofacial Disorders
Editors: Sh.B. Turner and J. E. Miller
ISBN: 978-1-62618-923-2

Chapter 1

Temporomandibular Joint Disorders: Etiopathogenesis, Classification, Diagnosis and Treatment

Alexandrina L. Dumitrescu[*1], *Viorel Cioranu-Ibric*[2] *and Sorin Ibric*[2]
[1]Private practice, Bucharest, Romania
[2]University "Lucian Blaga", Sibiu

Abstract

Temporomandibular joint (TMJ) disorders are a heterogeneous group of diseases that affects an increasing number of people, with 3-7% of the population seeking treatment for pain and dysfunction of the ATM. Apart from jaw pain, other common symptoms of TMJ disorders include limited or painful jaw movement, headache, neck pain or stiffness, clicking or grating within the joint, and, occasionally, an inability to open the mouth painlessly. The progressive joint degeneration leads finally to chronic pain and reduced quality of life. The etiology of TMJ disorders remains unclear, but it is likely multifactorial, with abnormal occlusion and emotional stress being considered the most serious etiological factors.

* Corresponding author: Alexandrina Lizica Dumitrescu, Email: alexandrina.l. dumitrescu@ gmail.com.

Other risk factors includes age, gender, hyperlaxity, genetic factors, orthodontic treatment and parafunctional habits such as bruxism, teeth clenching, lip biting. The correct diagnosis of TMJ dysfunctions cannot be based on clinical examination only and several methods of imaging can be recommended: X ray examination, computer tomography, magnetic resonance imaging, radionuclide examination (scintigraphy), ultrasonography and arthroscopy. For most patients, the signs and symptoms of TMJ disorders improve over time with or without treatment. Therefore self-care and noninvasive treatments (e.g. acupuncture, hypnosis, temporary occlusal therapy, intra-articular corticosteroid or anesthetic injection, pharmacologic treatment, physical therapy modalities and psychological interventions such as cognitive behavior therapy, relaxation techniques or stress management) are good options and should be attempted before invasive or permanent therapies, such as orthodontics or surgery, are recommended. This chapter reviews the current knowledge in the etiopathogenesis, classification, diagnosis of TMJ disorders and discusses several surgical and non-surgical treatment protocols.

I. Introduction

The Temporomandibular joint disorders (TMDs) form a cluster of related disorders with common symptoms which include localised pain, limited or asymmetric movement, and clicks or grating on opening (Mackie & Lyons, 2008). According to the glossary of prosthodontic terms, temporomandibular disorders (TMD) are defined as: "conditions producing abnormal, incomplete, or impaired function of the temporomandibular joint(s)", while medical dictionary defines temporomandibular joint (TMJ) disorder: "A disorder of the temporomandibular joint(s) that causes pain, usually in front of the ear(s), sometimes in the form of a headache".

Temporomandibular joint (TMJ) disorders negatively affect the Oral health-related quality of life of an increasing number of people. The impact appeared to be more pronounced in patients with more signs and symptoms (Dahlström & Carlsson, 2010).

II. Epidemiology of TMDs

Analyzing the prevalence of the signs and symptoms in TMD patients in epidemiological studies of TMD are important for establishing prevention and control programs due to the high level of knowledge of symptomatic complexes and therapeutic approaches that are revealed.

Table 1. Prevalence (%) of signs and symptoms of TMD in different epidemiological studies on adults (since 2000 to present)

Investigator	Country	Study population	Main findings
Bagis et al. (2012)	Turkey	243 (171 females, 72 males, mean age 41 years)	92%, pain in the temporal muscle; 89& pain during mouth opening
Poveda-Roda et al. (2012)	Spain	162 (mean age 40.6±18.8 years, 11.1% males and 88.9% females)	42% myofascial pain; 16% myofascial pain plus arthralgia
Han et al. (2011)	Korea	4,546 adults aged 18 years and older	15.3% TMD symptoms
Dallanora et al. (2012)	Brazil	Population of complete denture wearers	55.12%.
Plesh et al. (2011)	USA	189,992 people were included: 52% female and 48% male, 73% White, 12% Hispanic, 11% Black, and 4% "Other."	4.6% TMJ pain; 15.4% severe headaches/migraine; 14.9% neck pain; 28.0% low back pain
Balke et al. (2010)	Germany	223 subjects between (18 - 65 years of age)	20.2% - 46.2% facial pain
Gonçalves et al. (2010)	Brazil	1,230 subjects (51.5% women; 15 to 65 years old)	39.2% with at least one TMD symptom; 25.6% pain related toTMD
Gurbuz et al. (2010)	Turkey	118 patients with mental retardation; 104 mentally healthy individuals	69.2-79.7% at least one sign of TMD observed in patients with MR; 28.8 - 43.2% more than one sign of TMD
Machado et al. (2009)	Brazil	357 patients (mean age was 32 years)	86.8% of patients were women and 93.3% of the patients presented more than one diagnosis
Isong et al. (2008)	USA	30,978 people, 17,498 females and 13,480 males, 20,389 non-Hispanic whites and 4179 non-Hispanic blacks	4.6% overall TMD-type pain, with 6.3% for women and 2.8% for men
Nomura et al. (2007)	Brazil.	218 dentistry students (96 men and 122 women, with an average age of 20 years)	53.21% some level of TMD: 35.78% mild TMD 11.93% moderate and 5.5% severe
Cooper & Kleinberg (2007)	USA	4,528 patients, presenting over a period of 25 years to a single examiner for TMD treatment	96.1% pain; 79.3% headache; 75.0% temporomandibular joint discomfort or dysfunction; 82.4% ear discomfort or dysfunction; 85.1% tenderness to palpation of the pterygoid muscles; 62.4% tenderness to palpation of the temporomandibular joints
Marklund & Wanman (2007)	Sweden	371 1st year dental students	12% TMD signs and/or symptoms; 10% TMD sounds; 8% TMD pain

Table 1. (Continued)

Investigator	Country	Study population	Main findings
Uhac et al. (2006)	Croatia	94 Croatian war veterans suffering from post-traumatic stress disorder	63.83% experienced headache; 12.77% facial pain; 10.64% pain in the region of the jaw
Storm & Wänman, 2006	Sweden	487 Sami females	32% pain and/or dysfunction in the jaw-face region; 61% headaches; 56% pain in the cervical region
Rutkiewicz et al. (2006)	Finland	8,028 Finns at least 30 years of age	38% of the subjects had at least one sign of TMD.
Unell et al. (2006)	Sweden	6,343 subjects50-year-old	18-28% bruxism
de Oliveira et al. (2006)	Brazil	2,396 students (73% women and 27% men (mean age 21.94 4.8 years)	56.26 among men; 73.03% in women
Nekora-Azak et al. (2006)	Turkey	1253 individuals aged 18 years	31% jaw pain
Casanova-Rosado et al. (2006)	México	506 subjects aged 14-25 years	46.1% of the subjects exhibited some grade of TMD
Manfredini et al. (2004)	Italy	319 patients (222 females, 77.9%; 63 males, 22.1%; mean age 40 years old)	50.2% prevalence for Group I disorders (muscle disorders), 38.6% for Group II disorders (disc displacement), and 50.2% for Group III disorders (arthralgia, osteoarthritis, osteoarthrosis)
Miyake et al. (2004)	Japan	3557 Japanese university students, aged between 18 and 26 years	41.7% TMJ noise, 16% TMJ pain, 16.3% impaired mouth opening
Jagger et al. (2004)	UK	140 consecutive indigenous Indians (69 Quechua and 71 Colorado)	41% of at least one TMD symptom (higher prevalence of difficulty in opening of the mouth and pain in front of the ears) and 63% of at least one TMD sign
Gesch et al. (2004)	Germany	7,008 individuals (20 to 79 years old)	49.9% had one or more clinical signs of TMD, but only 2.7% were subjectively aware of TMJ pain symptoms; 5% tenderness to palpation; 15% masticatory muscle tenderness; 25% joint sounds; 9% limited maximum mouth opening < 40 mm; 1% pain upon movement of the mandibe; 28% irregular jaw movements (deviation, deflection)
Dulcić et al. (2003)	Croatia	96 subjects, 48 complete denture wearers and 48 removable partial denture wearers, whose average age was 75.8 years	9.3% reported symptoms of internal derangement of TMJ; tissue-specific diagnoses were established in 52.1% of the subjects
Dulcić et al. (2003)	Croatia	196 partially edentulous subjects (Eichner's class II	16.3% had symptoms and signs of TMD; 46.9% of the subjects a tissue-specific

Investigator	Country	Study population	Main findings
		and III) at the average age of 51.2.	diagnosis of TMD
Nassif et al. (2003)	Saudi Arabia	523 young adult males	75% of the subjects had TMD symptoms and/or signs. (6.9% insignificant moderate; 51.4% moderate; 16.7% severe)
Yap et al. (2003)	China	191 (53 male and 138 female; mean age 33.6 +/- 9.3 years)	31.4% muscle disorders; 15.1% and 15.7% disc displacement in the left and right joints; 12.6% and 13.0% presented arthralgia, arthritis, and arthrosis in the left and right joints
Katz and Heft (2002)	Israel	20,869 young soldiers aged 18-21 years	8.4 percent (8.9% in females and 8.2% in males) TMJ sounds; 4% masticatory pain
Choi et al. (2002)	Korea	27,978 Nineteen-year old men	17.8% masticatory muscle stiffness; 14.3% TMJ sounds; 7.2% headache; 13.5% neck pain; 8.4% bruxism; 9.9% clenching
Magnusson et al. (2002)	Sweden	114 subjects was followed longitudinally from the age of 15 to 35 years	The estimated need for treatment was 11%, but the active demand for such treatment was much less (3%)
Pow et al. (2001)	Hong Kong	1,526 randomly selected Cantonese-speaking individuals aged 18 years or over	33% jaw pain (5% with frequent pain); 0.3% jaw opening; 1.8%, joint clicking
Sönmez et al. (2001)	Turkey	182 children with mixed dentition and 212 with permanent dentition	Prevalence of TMJ dysfunction was 68% in mixed dentition and 58% in permanent dentition
Magnusson et al. (2000)	Sweden	135 randomly selected 15-year-old subjects	13% reported 1 or more frequent TMD symptoms
Otuyemi et al. (2000)	Nigeria	308 medical and dental students (207 males and 101 females; mean age of 23 +/- 3.0 years)	26.3% mild TMD symptoms; 2.9% severe TMD symptoms; 46% mild, 16.5% moderate and 0.3% severe TMD signs

II. (1) Prevalence of Temporomandibular Pain in Adults

Several surveys have been conducted all over the world. As can be seen in Table 1, the prevalence rates reported for TMD have varied widely, ranging from 6% to 93% , less likely due to variations in the populations studied, but most probably due to different clinical criteria used in the mentioned studies to define TMD (Poveda Roda et al., 2007; Manfredini et al., 2012). Most research groups have focused on pain in the structures of the temporomandibular region as the defining characteristic of a temporomandibular disorder (LeResche, 1997).

In contrast to what might be expected, there is a poor correlation between the radiographic imaging findings in relation to corresponding clinical findings. When radiographic changes of TMJ disorders were investigated, the prevalence of bone change in the condyle correlated poorly with age, sex, and dental and occlusal condition. Deformity was the most common abnormality of bone change for TMD patients, followed by the high rate of osteophytes (Takayama et al., 2008).

Based on the TMD quantification according to the population demands for treatment, only 3-7% of the population seeks treatment for pain and/or dysfunction of the TMJ or related structures (Carlsson, 1999). Magnusson et al. (2012) has followed longitudinally from the age of 15 to 35 years an epidemiological sample of 114 subjects, with regard to treatment received, treatment demand, and treatment need for TMD. During the 20-year follow-up period, the estimated need for treatment was 11%. Only 3% still had a demand for such treatment at age 35. This reduced prevalence rate can be explained that those patients who do not seek treatment do not consider such alterations to be a relevant problem and are able to lead a normal life despite the symptoms (Poveda Roda et al., 2007; Magnusson et al., 2002; Carlsson, 1999). Most of those who seek treatment for TMD are female (female-male ratio being at least four to one) and patients most commonly in early adulthood (Dimitroulis, 1998).

II. (2) Prevalence of Temporomandibular Pain in Children

Information on the patterns of such temporomandibular pain in children is necessary to complete the epidemiologic picture across the life cycle (LeResche, 1997). Several studies that had investigated the prevalence of pain in the face or jaws in children, adolescents, and young adults were identified and presented in Table 2. Prevalence rates of pain in the temporomandibular region found in these investigations ranged from 2.7-68%. Moyaho-Bernal et al. (2010) reported a high prevalence of signs and symptoms related to temporomandibular disorders, in particular in children with mixed dentition, 33.2% predominately muscular (82%), with signs of muscular pain in 48.1%, joint pain in 19.1%, with signs of alteration in the mouth opening pattern in 63.4%, joint sounds on opening or closing the mouth in 39.1% presented and 20.4% on mandibular excursions. These results reveals shows the importance of TMD evaluation during the period of mixed dentition, when morphological changes associated to growth and craniofacial development prevail.

Table 2. Prevalence (%) of signs and symptoms of TMD in different epidemiological studies on children

Investigator	Country	Study population	Main findings
Wu and Hirsch (2010)	Germany	1,058 subjects (561 German, 497 Chinese) aged 13 to 18 years	13.9% prevalence of TMD diagnoses
Moyaho-Bernal et al. (2010)	México	235 children (54.9% female and 45.1% male, of average age 9.31 + 1.2 years)	33.2% signs and symptoms; 82% muscular signs, 48.1% muscular pain; 19.1% joint pain; 63.4% alteration in the mouth opening pattern; 39.1% joint sounds on opening or closing the mouth; 20.4% joint sounds on mandibular excursions
LeResche et al. (2007)	USA	1,996 boys and girls, initially 11 years old	6.8% TMD pain
Feteih (2006)	Saudi Arabia	385 (230 females and 155 males) school children age 12-16	21.3% TMD signs with joint sounds being the most prevalent sign; 33% TMD symptoms with headache being the most prevalent; 41% lip/oral parafunctions with 29% cheek biting followed by nail biting
Nilsson et al. (2005)	Sweden	28,899 youths aged 12 to 19 years	4.2% TMD pain. The prevalence increased with age, and a significant difference was seen between boys (2.7%) and girls (6.0%)
Tuerlings and Limme (2004)	Belgium	136 children (70 boys, 66 girls) aged from 6 to 12 years	80.9% lateral pterygoid muscle tenderness; 35.3% TMJ sounds; 19.8% presented a mandibular deviation on maximal opening; 57.4% retruded contact position interferences; 72.1% lateral and protrusive interferences
Sönmez et al. (2001)	Turkey	182 children with mixed dentition and 212 with permanent dentition	Prevalence of TMJ dysfunction was 68% in mixed dentition and 58% in permanent dentition

An increase in signs and symptoms from the primary to the mixed dentition group was reported (Muhtaroğullari et al., 2004) and the total prevalence of signs and symptoms of TMJ dysfunction in mixed dentition was found to be higher than in permanent dentition. As Sönmez et al. (2001) reported 68% (68% in girls and 68% in boys) in mixed dentition and 58% (61% in girls and 56% in boys) in permanent dentition. The gender differences in prevalence seen in adults are not apparent in children as no statistically significant difference was found in the total prevalence of TMJ dysfunction between girls and boys (Sönmez et al., 2001; Ogura et al., 1985).

In a recent review, Howard (2013) emphasized the increasing the importance of the dentist's awareness of the early signs and symptoms of TMD, as child's difficulty in verbalizing the precise location and nature of facial pain and jaw dysfunction often results in a nondefinitive history, while an accurate diagnosis enables timely referral to appropriate health care providers.

III. Classification of TMDs

Presently, there are several main classifications related to TMD:

1. the Research Diagnostic Criteria (RDC) for TMD (Dworkin & LeResche, 1992)
2. the Wilkes Classification for TMJ internal derangement (Wiles, 1989)
3. the American Academy of Orofacial Pain (AAOFP) Classification of TMD (de Leeuw, 2008)

Table 3. Research diagnostic criteria for temporomandibular disorders (RDC/TMD) version 1.0 Axis I diagnoses.

Group I Muscle Disorders:	o (Ia) myofascial pain; o (Ib) myofascial pain with limited opening.
Group II Disc Displacements:	o IIa) disc displacement with reduction; o (IIb) disc displacement without reduction with limited opening; o (IIc) disc displacement without reduction without limited opening.
Group III Arthralgia, Arthritis, Arthrosis:	o (IIIa) arthralgia; o (IIIb) osteoarthritis; o (IIIc) osteoarthrosis

The Research Diagnostic Criteria for Temporomandibular Disorders (RDC/TMD), proposed in 1992, is a standardized and a well-operationalized diagnostic scheme that has been widely used in epidemiological, clinical and experimental studies. The RDC/TMD is based on two diagnostic axes:

- Axis I establishes a diagnosis based on three clinical variables (myofascial pain, disc alterations and arthralgia-arthritis-arthrosis), each with several subcategories. As a consequence, patients can be assignd to any of 3 diagnostic groups that include 8 sub diagnoses (Table 3).
- Axis II establishes a diagnosis based on psychological variables.

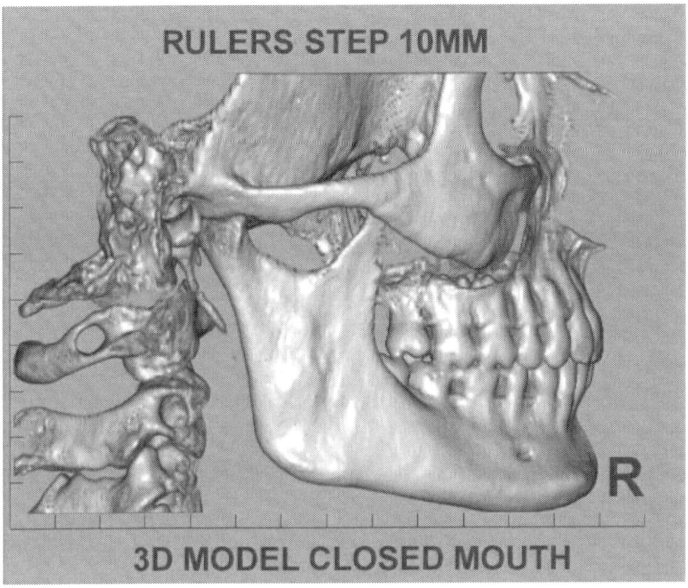

Figure 1. Tridimensional CT reconstruction of a TMJ with osteoarthritis.

The RDC/TMD Axis I diagnostic algorithms for clinical TMD diagnoses have been recently evaluated. It was demonstrated a high sensitivity and specificity for myofascial pain (0.82, 0.99, respectively), myofascial pain with limited opening (0.93, 0.97), combining diagnoses for any myofascial pain (0.91, 1.00, respectively), joint pain when arthralgia and osteoarthritis (Figure 1.) were combined (0.92, 0.96 respectively) and for disc displacement without reduction with limited opening (0.80, 0.97, respectively). However, for the other disc displacement diagnoses, osteoarthritis and osteoarthrosis, revised clinical criteria alone, without recourse to imaging, are inadequate (Schiffman et al., 2010). Reliability of the RDC/MD Axis I protocol was also investigated. Intersite reliability of RDC/TMD diagnoses was good to excellent for myofascial pain, arthralgia, disc displacement with reduction, and disc displacement without reduction with limited opening. Reliability was poor to marginally fair for disc displacement without reduction without limited opening and osteoarthrosis (Look et al., 2010).

The prevalence of different RDC/TMD axis I diagnoses in TMD patients and in the general populations was recently summarized and systematically reviewed by Manfredini et al. (2012). Muscle disorders, disk displacements, and other joint disorders were diagnosed respectively in 56.4%, 42.0%, and 57.5% of five hundred twenty consecutive patients seeking TMD treatment.

While Axis I permits the diagnosis based on three common forms of muscle and joint involvement, Axis II establishes a diagnosis based on psychological variables, such as psychological distress and psychological dysfunction (Cavalcanti et al., 2010). RDC/TMD Axis II Questionnaire consists of 31 items, divided into (de Lucena et al., 2006):

- socio-demographic,
- socio-economic,
- psychological (depression subscales and nonspecific physical symptoms – pain items included and pain items excluded)(e.g. Feeling low in energy, Thoughts of dying, Crying easily, Poor appetite, Blaming yourself for things, Feeling no interest in things, Feeling blue, Feeling hopeless about future, Worrying too much about things)
- psychosocial (graded chronic pain severity – pain intensity and disability);
- patient-related signs and symptoms and the limitation scale on mandibular function (limitations related to mandibular functioning) (e.g Chewing, Drinking, Exercising, Eating hard foods, Eating soft foods, Smiling/laughing, Cleaning teeth or face, Swallowing, Talking)

The Axis-II Depression and Graded Chronic Pain instruments demonstrate psychometric properties suitable for comprehensive assessment and management of TMD patients. The Axis II instruments showed very good-excellent internal consistency (Cronbach's alpha = 0.80 – 0.95), convergent (correlation range 0.3–0.9) and discriminant (range 0.0–0.6) validity and temporal stability (Lin's correlation concordance coefficient [CCC] = 0.91)(Ohrbach et al., 2010; Dworkin et al., 2002)

When the prevalence of different RDC/TMD axis II diagnoses in TMD patients was systematically reviewed, it was revealed that 60% of patients had depression symptoms, 76.6% had somatization, and 21.8% presented high levels of pain-related impairment (Manfredini et al., 2012).

One widely used classification, adopted by surgeons who treat TMJ disorders, is the Wilkes classification. It describes escalating joint pathology in 5 stages (early, early/intermediate, intermediate, intermediate/late, late) focusing mainly on internal derangement and osteoarthritis (Wilkes, 1989).

Table 4. Criteria for new TMJ surgical classification proposed by Dimitroulis (2013)

Category	TMD	Diagnosis	Evaluation	Surgical treatment
1	TMJ normal	Normal	Normal	None
2	TMJ minor changes	TMJ early internal derangement Joint inflammation/adhesions	All joint components are salvageable	TMJ arthrocentesis TMJ arthroscopic lavage
3	TMJ moderate changes	TMJ moderate internal derangement TMJ recurrent dislocation TMJ synovial chondromatosis Dislocated condylar fracture	Most joint components are salvageable	TMJ arthroscopy (operative) TMJ arthroplasty – disc plication/repositioning Modified condylotomy
4	TMJ severe changes	TMJ discectomy _ condyloplasty/shave Debridement of glenoid fossa Reduction of eminence	Few joint components are salvageable	TMJ discectomy ± condylar surgery Debridement of glenoid fossa Reduction of eminence
5	TMJ catastrophic changes	TMJ osteoarthritis TMJ condylysis TMJ ankylosis TMJ tumour	Nothing in the joint is salvageable	TMJ resection TMJ total joint replacement

The Guidelines of the American Academy of Orofacial Pain from 2008 include in TMJ disorders the followings: congenital or developmental disorders, disc derangement disorders, TMJ dislocation, inflammatory disorders, non-inflammatory disorders, ankylosis, and fracture. Masticatory muscle disorders include local myalgia, myofascial pain, centrally mediated myalgia, myospasm, myositis, myofibrotic contracture, and masticatory muscle neoplasia (de Leeuw, 2008). However, one problem with the American Academy of Orofacial Pain classification of TMD is related to data collection as it does not allow for degree of disability which is quantifiable (Dimitroulis, 2013).

Regarding TMJ dislocation, Akinbami (2011) proposed a classification based on relationship of the head of mandibular condyle to the articular eminence seen on clinico-radiological evaluation into three types (I-III): Type I - the head of condyle is directly below the tip of the eminence; Type II - the head of condyle is in front of the tip of the eminence; Type III - the head of condyle is high up in front of the base of the eminence.

In aim to collate the evidence required to justify the surgical interventions undertaken to treat TMJ disorders, Dimitroulis (2013) recently proposed a new classification that divides TMJ disorders into five categories of escalating degrees of joint disease, with category 1 being normal and category 5 referring to catastrophic changes to the joint, classification that can be applied to TMJ surgery research and clinical practice (Table 4.).

IV. Etiopathogenesis of TMDs

The etiology of TMJ disorders remains unclear, but it is likely multifactorial, involving a large number of direct and indirect causal factors (Ibric Cioranu et al., 1999):

1. Age

In adults, age (odds ratio 0.954; 95% CI 0.922-0.987, $P = 0.006$) had significant effects on the prevalence of TMD (Bagis et al., 2012). Similarly, late adolescent (16- to 18-year-olds) patients with TMDs have higher pain intensity in the orofacial region than do early adolescent (13- to 15-year-old) and juvenile patients (6- to 12-year-olds) with TMDs (Karibe et al., 2012).

2. Gender

Sex-specific differences in temporomandibular disorders have been observed with TMJ pain at rest, pain in the masseter muscle, muscle tenderness on palpation clicking, crepitation, and bruxism being significantly more frequent in females than males (Schmid-Schwap et al., 2013; Bagis et al., 2012; Plesh et al., 2011). This might be related to biological, anatomical, or hormonal factors acting alone or in combination in females (Bagis et al., 2012). No gender differences were noted in children, adolescents and young students (Karibe et al., 2012; Bonjardim et al., 2009; Nilsson et al., 2009).

3. Ethnicity

Significant racial/ethnic differences were revealed for TMD-type pain and neck pain, with Whites being more likely to report these pains than either Blacks or Hispanics (Plesh et al., 2011; Plesh et al., 2002; Riley et al., 2002; Gansky & Plesh, 2007). However, while significantly higher rates were reported by Whites comparing with either of the other groups, the degree of difference between Blacks and Hispanics was smaller (Plesh et al., 2011).

4. Genetic Factors

An increasing number of studies have revealed the association between TMD and genetic polymorphisms in the oestrogen receptor alpha, adrenergic receptor beta 2, serotonin receptor, serotonin transporter and catechol-O-methyltransferase genes (Meloto et al., 2011; Oakley & Vieira, 2008). As a consequence a possibility of "genome/omics-based personalised care" was proposed by Meloto et al. (2011): depending on the individual genetic differences and early diagnosis and prognosis of the disorder, the individual treatment is based on personalised medication, preventing acute pain conditions from becoming chronic.

5. Smoking

Smokers with TMD reported higher pain severity than nonsmokers with TMD. Also these patients are at higher risk for factors that may adversely affect treatment outcomes (de Leeuw R et al., 2013; Sanders & Slade, 2011; Sanders et al., 2012; Weingarten et al., 2009). In univariate analysis, tobacco users were more likely to have moderate to severe pain interference defined as chronic Pain Grading Scale III-IV (OR=1.94, 95% CI 1.16-3.22) compared to non-users (Weingarten et al., 2009).

Also the effect of cigarette smoking on chronic pain is stronger in younger than older adults: compared with nonsmokers, ever smokers aged <30 years had higher odds of TMD (OR = 4.14, 95% CI: 1.57, 11.35) than older adults (OR = 1.23, 95% CI: 0.55, 2.78) ($P < 0.05$) (Sanders et al., 2012).

6. Occlusion

Different studies have identified alterations in occlusion such as Angle malocclusions, crossbite, open bite, occlusal interferences, prominent overjet and overbite, class II and distal molar occlusions crowding, midline discrepancies and missing teeth as predisposing, triggering or perpetuating factor of TMD (Poveda Roda et al., 2007; Mohlin & Thilander, 1984;

Egermark-Eriksson et al., 1983). There are indications that loss of posterior support may be a contributing factor for TMJ pain (Ciancanglini et al, 1999; Ciancanglini et al, 2003; Seedorf et al, 2004).

Table 5. Recent articles who systematically reviewed how occlusion and malocclusion may relate to TMD

Author, year	Nr. of studies included	Main conclusion
Luther (2007a)	28 citations	Evidence is lacking to suggest static occlusal factors cause TMD.
Luther (2007b)	40 citations	Neither static nor dynamic occlusal factors (including orthodontics) can be said to "cause" TMD.
Gesch et al. (2004)	22 citations	Few associations were established between malocclusion or functional occlusion and signs and symptoms of TMD
Mohlin et al. (2007)	Literature search from 1966 to 2005	Associations between specific types of malocclusions and development of significant signs and symptoms of TMD could not be verified.
Thilander & Bjerklin (2012)	14 citations	A functional posterior crossbite (mandibular guidance with midline deviation) is associated with headache, temporomandibular joint and muscular pain, and clicking

Unfortunately, it was showed that these studies are open to substantial criticism as they are have generally not been carried out in an ideal manner, vast majority of studies in the orthodontic literature have used an inappropriate index such as Helkimo Index or a modification of it, the investigators were not blinded, the authors were usually orthodontists, while the correlations that have been reported between TMD and the various malocclusion types are low and unlikely to be of direct clinical significance (even if statistically significant) (Luther, 2007a; Türp & Schindler, 2012). Insertion of an oral splint or occlusal adjustment could demonstrate a symptom improvement but it was not able to prove an occlusal aetiology because the amelioration may be due to the change of the appliance-induced intermaxillary relationship (Türp & Schindler, 2012). In addition, symptoms often abate even in the absence of therapy. In this line, a number of recent systematic reviews failed to reveal an association between malocclusion or functional occlusion and signs and symptoms of TMD (**Table 5**). It becomes apparent that the evidence of a causal relationship between occlusion and temporomandibular disorders is weak. Only bruxism, loss of posterior support and unilateral posterior crossbite show some consistency across studies (Türp & Schindler, 2012).

7. Orthodontic Treatment

The possibility that orthodontic treatment could cause TMJ pathology has been extensively investigated in the scientific literature in: a) Treated *versus* untreated patients, b) Patients treated by different means and c) in Longitudinal studies - the same patients are compared before and after treatment (Luther, 2007a). However, most studies were not prospective, were undertaken without using a reproducible and validated index, were not blinded, randomized, appropriately powered or using appropriate controls. Based on the available evidence, the widely accepted conclusion is that orthodontic treatment neither causes nor cures TMD (Luther, 2007a; Luther et al., 2010). However, one hypothesis of causality has been proposed for studies who positively found a correlation: because orthodontic treatment lasts quite long time, around 2 years, orthodontic patients may complain about TMD during or after treatment (Michelotti & Iodice, 2010).

8. Parafunctional Habits Such as Bruxism, Teeth Clenching, Lip Biting

It has been reported that bruxism is a risk factor for masticatory muscle pain (Velly et al, 2003). The mechanisms by which it is thought that parafunctional habits cause TMD is either microtrauma or masticatory muscle hyperactivity that can became self-perpetuating and could lead to organic disease. However, these habits are also common in asymptomatic patients (Buescher, 2007).

Two papers systematically reviewed the literature on the TMD-bruxism relationship in adults from 1998 to 2008 (Manfredini & Lobbezoo, 2010) and in children and adolescents from 1970 to 2007 (Barbosa Tde et al., 2008). In adults, self-report or clinical bruxism diagnosis showed a positive association with TMD pain, association affected by some potential bias and confounders (eg, pain as a criterion for bruxism diagnosis), while quantitative and specific methods to diagnose bruxism showed much lower association with TMD symptoms (Manfredini & Lobbezoo, 2010). Similar results were found in children and adolescents, were the relationship between bruxism and TMD seems to be controversial and unclear, not only because some studies have shown and association between TMD to bruxism in the primary and mixed dentition, whereas other authors did not observed respective relationship in primary dentition, but also due to the unreliability for the clinical assessment of bruxism in the investigated literature (Barbosa Tde et al., 2008).

9. Joint Hyperlaxity

Both local and systemic hyperlaxity has been postulated as a possible cause of TMD. In a literature review for the association between TMD and generalized joint hypermobility (3 studies included, 113 cases and 95 controls) the odds ratio was 5.4 (26 cases and 5 hypermobile controls). However, in a sensitivity analysis, the odds ratio changed from significant to non-significant (Dijkstra et al., 2002).

These results are in contrast with two recent studies. A retrospective study of a series of 850 patients with temporomandibular dysfunction showed that ligament hyperlaxity was one of the variables showing statistically significant differences among the four diagnostic categories: temporomandibular dysfunction-myofascial pain, disc displacement with reduction, disc displacement without reduction and osteoarthrosis (Poveda-Rona et al., 2009). In a case-control study, Poveda-Rona et al. (2012) compared the risk factors and clinical manifestations of patients with TMDs (mean age 40.6±18.8 years, range 7-90; 11.1% males and 88.9% females) diagnosed according to the Research Diagnostic Criteria for Temporomandibular Disorders (RDC/TMD) (axis I) versus an age and gender matched control group (162 cases vs 116 controls). The ligament laxity showed no statistically significant differences between patients and controls, though the observed tendency could be of clinical relevance ($P = 0.09$). However, myofascial pain and ligament laxity were related with an odds ratio 2.6 (CI: 1.01-6.68).

10. Psychosocial Risk Factors for Chronic TMD

Psychological factors have been closely linked to TMD. Higher levels of somatization, distress, anxiety, depression and neuroticism, and minor use of positive reinterpretation and humor, as coping strategies, affect TMD sufferers negatively compared to people without TMD (Ferrando et al., 2004; Suvinen et al., 2005; Eli, 2003; Manfredini et al., 2003). For example, in a prospective cohort study of healthy female volunteers aged 18-34 yrs, depression, perceived stress, and mood were associated with pain sensitivity and were predictive of 2- to 3-fold increases in risk of TMD ($P < 0.05$)(Slade et al., 2007).

V. Clinical Examination

The Council of the European Academy of Craniomandibular Disorders charged the Educational Committee with the task to establish Guidelines and Recommendations for the examination, diagnosis and management of patients with temporomandibular disorders and orofacial pain by the general dental practitioner (De Boever et al., 2008).

First, a routine screening protocol is recommended in order to detect the presence of a TMD. The protocol includes four (mostly validated) questions regarding: 1) presence, once a week or more, of pain when open your mouth wide or chew; 2) presence, once a week or more, of pain in temples, face, temporomandibular joint or jaws; 3) presence of locked jaw that cannot be opened wide; 4) presence, more than once a week, of headache. A positive answer to one of the 4 questions reveals reveals the necessity of a TMJ history taking and assessment (De Boever et al., 2008).

There are three cardinal features of TMD (Ohrbach and Gale, 1989; Dworkin et al., 1990; Dolwick, 1989; Green and Laskin, 1988; Dimitroulis, 1998):

1. orofacial pain (the most common presenting complaint and the most difficult problem to evaluate),
2. joint noise (common in asymptomatic people in the general population, and less important clinically in the absence of pain) and
3. restricted jaw function (associated with anxiety and difficulties in eating and speaking due to the limitation of mandibular movements)

Clinical examination of the TMD patient includes (De Boever et al., 2008):

- assessment of the contributing factors involved, and their grading;
- extra-oral inspection of head and neck areas (asymmetries, posture, scars) as well as TMJ region (Figure 2.). Areas of tenderness, trigger points, and patterns of pain in muscles and TMJ should be noted. TMJ

palpation is performed by placing the finger tips in the preauricular region just in front of the tragus of the ear. When the patient is opening his mouth, the finger tip will fall into the depression left by the translating condyle. Joint sounds and their location during opening, closing, and lateral excursions of the mandible may be either palpated or detected with a stethoscope placed over the preauricular area.

- intra-oral inspection and examination including occlusal characteristics and assessment of parafunctions (e.g. bruxism);
- physical examination of the masticatory system (e.g. digital palpation of the masticatory musculature). Evaluation of the mandibular function can be made by noting whether the line of vertical opening is straight and smooth or deviates with jerky movements. It is also necessary to record so the range of painless maximal vertical opening (normal range 42-55 mm interincisal distance);
- assessment of psychosocial factors according to a validated checklist (Axis II evaluation);
- presence of sleep disorders or of systemic diseases, case in which the patient should be referred via the general medical practitioner to the appropriate specialist.

VI. Complementary Explorations

Often clinical examination isn't sufficient for a correct diagnosis of TMJ dysfunctions and, in this case, can be recommended several methods of imaging (e.g. computer tomography, magnetic resonance imaging), arthroscopy or biochemical analysis of the synovial fluid.

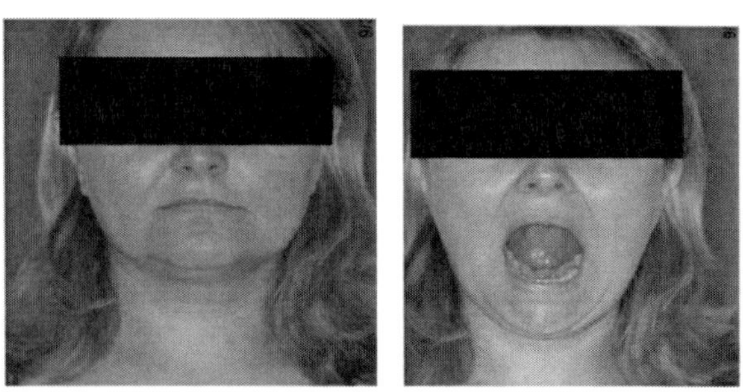

Figure 2. Extra-oral examination of a patient with an acute disk displacement.

VI.1. Imaging Techniques

Additional imaging techniques can be recommended when fractures, tumors and other hard or soft tissue lesions are suspected.

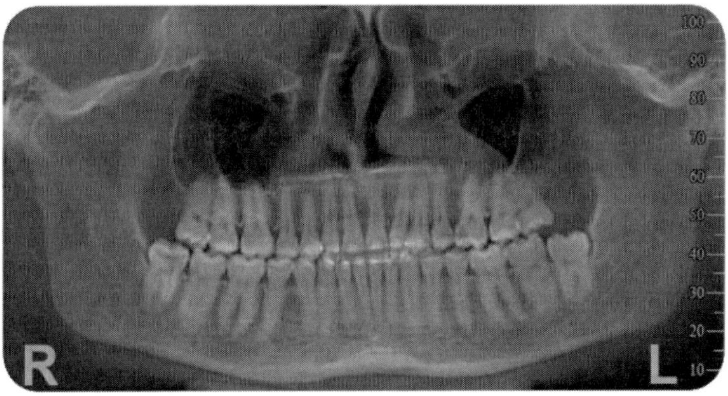

Figure 3. Panoramic radiography of a patient with treated osteorthrosis.

Panoramic X-rays

Panoramic radiography is routinely used in dentistry because it provides general information about teeth, mandible, and other regions of the jaw, but his use for TMJ imaging is still controversial, because of his magnification differences and image distortions (due to nonlinear variation in the magnification at the different object depths) as well its inability to visualize disk, joint space, and fossa. However, the dental panoramic radiography is a good technique to visualize only the bony components of the TMJ and the ascending rami in the mandible and the condyle (only the lateral and medial thirds of the condylar head) (Ahn et al., 2006; Habets et al., 1987; Ruf & Pancherz, 1995)(Figure 3.).

Computed Tomography (CT)

Several studies have documented the superiority of CT over panoramic radiography or MRI in displaying the features of TMJ (Honey et al., 2007; Westesson et al., 1987; Hussain et al., 2008). Hussain et al. (2008) indicated that axially corrected sagittal tomography is currently the imaging modality of choice for diagnosing erosions and osteophytes in the TMJ and that high-resolution ultrasound and cone beam CT may be superior to axially corrected sagittal tomography due to cost- and radiation dose-effective capacity. Ahmad et al. (2009) revealed that that about 75% of CT-diagnosed osteoarthritis are not detected using panoramic radiography, and about 40% with MRI, suggesting that CT should be used when possible despite the increased radiation. However, radiation exposure can be decreased by limiting CT study to the closed-mouth position or by using a cone beam CT, which can provide diagnostic information equal to multidetector CT.

Cone beam CT has been shown to provide high-resolution multiplanar images by delivering substantially lower radiation dose. Cone beam CT represents a better diagnostic option for TMD as allows examination of TMJ anatomy without superimposition and distortion to facilitate analysis of bone morphology, joint space and dynamic function in all three dimensions (Honey et al., 2007; Barghan et al., 2012; Barghan et al., 2010)(Figure 4.).

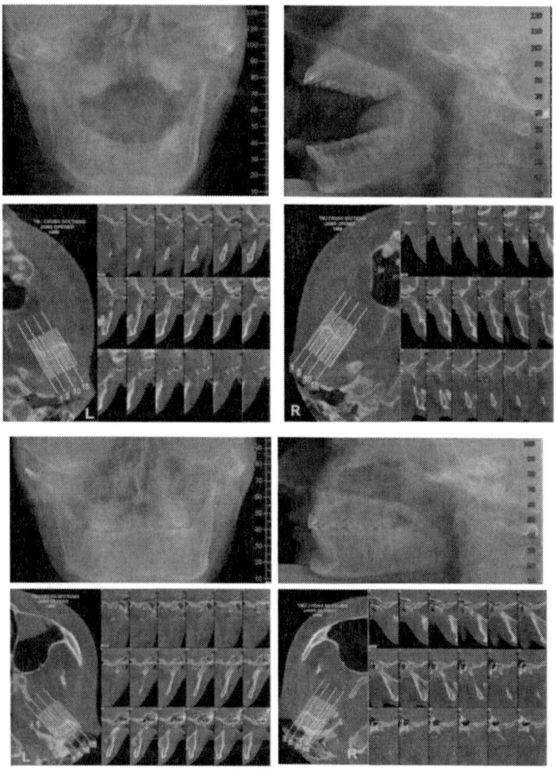

A.

B.

Figure 4. Computer tomography of a patient with treated TMJ osteoarthritis (A. open mouth; B. closed mouth).

Magnetic Resonance Imaging (MRI)

MRI permits a detailed assessment not of the TMJ anatomy due to its inherent tissue contrast and high resolution, of the TMJ biomechanics through imaging patients in the closed and open jaw positions and of TMJ disorders such as diagnosis of disc position and configuration, disc perforation, joint effusion, and osseous and bone marrow changes in the TMJ (Ribeiro-Rotta et al, 2011). Interpretation of MR imaging requires knowledge of the normal anatomy and an understanding of normal and abnormal biomechanics (Aiken et al., 2012). Compared with arthrography and CT, MRI allows a more correct view of the degenerative bone changes (60–100 %), disc dislocation (73–95 %) as well as synovial and subsynovial tissue (Tomas, 2006).

Two systematic reviews (Koh et al., 2009; Ribeiro-Rotta et al, 2011) evaluated the relationship between diagnoses and findings of clinical examination and diagnoses and findings of MRI examination for degenerative and inflammatory temporomandibular joint diseases but no clear evidence was found for a relationship between clinical and MRI diagnoses and findings. The calculated odds ratio (OR) for the relationship between clinical pain and internal derangements diagnosed with MRI was generally low (1.54-2.04), while the ORs for the relationship between pain and disc displacement without reduction (4.82) or between crepitation and disc displacement without reduction (3.71) were higher (Koh et al., 2009). MRI revealed a fair reliability for the radiologic diagnosis of TMJ ostheoarthritis (kappa statistic = 0.46), but excellent for diagnosing disc displacements with reduction (kappa statistic = 0.78) and for DD without reduction (kappa statistic = 0.94), while the overall percent agreement for pairwise ratings was 59% for diagnosing osteoarthritis but 95% for any disc displacements (Ahmad et al., 2009).

Ultrasonography

Ultrasonography (US) has been shown to be an important diagnostic tool, comparable to magnetic resonance imaging (MRI), for the assessment of TMJ involvement. Better visualization of joint structures and more reliable results with higher sensitivity and accuracy can be achieved with high-resolution devices (≥12 MHz)(Bas et al., 2011).

US proved to be accurate in detecting normal disc position and the presence of abnormalities in disc-condyle relationship (sensitivity of 65.8%, specificity of 80.4%, diagnostic odds ratio of 7.97), but not so useful for the distinction between disc displacement with and without reduction (Tognini et al., 2005). US imaging revealed a very good diagnostic accuracy for TMJ effusion (sensitivity of 75.6%, specificity of 76.5%, positive predictive values

73.8%, negative predective values 78.2%). US vs MRI agreement for the diagnosis of TMJ effusion was fairly good (pct. agreement 76.1%; K=0.521)(Tognini et al., 2002). US technique proved to be a valuable diagnostic aid for the detection and prediction of abnormal condylar morphology.

Table 6. Recent articles who systematically reviewed the diagnostic efficacy of ultrasonography in detecting TMDs

Author, year	Nr. of studies included	Main conclusion
Li et al., 2012	15 studies	The diagnostic efficacy of US is acceptable (diagnostic odds ratio for disc displacement with reduction was 10.95, while for disc displacement without reduction was 36.80) and can be used as a rapid preliminary diagnostic method to exclude some clinical suspicions. However, positive ultrasonographic findings should be confirmed by MRI
Manfredini & Guarda-Nardini, 2009	17 studies	US accuracy was 54-100% for diagnosing TMJ disk displacement, 72-95% for TMJ effusion and 56-93% for TMJ osteoarthrosis.
Melis et al., 2007	17 studies	US accuracy ranged from 51.5-100% for the evaluation of TMJ disk displacement, from 72.2-95% for the evaluation of TMJ effusion, and from 67-94% for the evaluation of condylar erosion.

The diagnostic accuracy results to assess the presence or absence of TMJ condylar osteoarthrosis: sensitivity of 87%, specificity of 20%, positive predictive value of 88%, negative predictive value of 18%, indicate that US may be valuable in diagnosing the presence, but insufficient in diagnosing the absence of osteoarthrosis (Brandlmaier et al., 2003). US may provide valuable information also about disk displacement of the TM (the accuracy of of internal derangement, disk displacement with reduction, and disk displacement without reduction was 95%, 92%, and 90%, respectively)(sensitivity of 64%, specificity of 88%, positive predictive value of 84%, and a negative predictive value of 71%, with an accuracy of 76%) (Emshoff et al., 2002; Byahatti et al., 2010). Several systematic reviews, summarized in Table 6, have the diagnostic efficacy of ultrasonography in detecting TMDs.

Single Photon Emission Computed Tomography

Single photon emission computed tomography (SPECT) of the bone is frequently performed in routine nuclear medicine practice, due to increased image contrast and improved lesion detection (Sarikaya et al., 2001). Radionuclide *99mTc* is used for the examination. Three-dimensional image analysis often aids lesion localization while its ability to detect functional changes offers more information before bone structural changes visible on radiographs occur (Güner et al., 2003).

Studies have documented the unique diagnostic information provided by SPECT, particularly for the diagnosis of temporomandibular joint internal derangement (Coutinho et al., 2006) and diagnosis of condylar hyperplasia (Villanueva-Alcojol et al., 2011; Saridin et al., 2009; Saridin et al., 2008). For patients with unilateral condylar hyperactivity, direct comparison of bone activity between the affected and contralateral condyle in SPECT scans is the analysis method of choice, determining the percentile bone activity in both condyles with a sensitivity of 93% and a specificity of 96% (Saridin et al., 2009; Saridin et al., 2008). Also in patients with a clinically suspected active unilateral condylar hyperactivity, SPECT scanning is the preferred diagnostic tool rather than planar bone scintigraphy (sensitivity of 67% and a specificity of 85%)(Saridin et al., 2008).

When the SPECT with Technetium 99m and computed tomography (CT) were evaluated during simultaneously image aquire in diagnosis of TMJ dysfunction, it was reported that the correlation of signs and symptoms with SPECT/CT imaging showed sensitivity 100%, specificity 90.91%, and accuracy 96.97% (Coutinho et al., 2006).

VI.2. Arthroscopy

Arthroscopy is both a therapeutic and mini invasive endoscopic method for TMJ space examination. TMJ arthroscopy facilitates joint space visual control in diagnosing and treating internal TMJ derangement, TMJ degenerative disorders, synovitis, TMJ hypo and hypermobility (Tvrdy, 2007).

The assessment of the effectiveness of arthroscopy for the management of signs and symptoms in patients with TMDs, revealed that 1) both arthroscopy and nonsurgical treatments reduced pain after 6 months; 2) open surgery was more effective than arthroscopy at reducing pain after 12 months but no differences in mandibular functionality or in other outcomes in clinical evaluations were noted; 3) arthroscopy led to greater improvement in maximum interincisal opening after 12 months than arthrocentesis; however, there was no difference in pain (Rigon et al., 2011).

VI.3. Biochemistry of Synovial Fluid

In aim to better understand the pathologic mechanisms of the TMDs, several studies have evaluated the chemical composition of the fluid that surrounds TMJ. Scientists have detected certain compounds (lubricin, of interleukin 6, vascular endothelial growth factor, interleukin-1beta, tumor necrosis factor alpha, matrix metalloproteinase-2, substance P, N-acetyl-beta-glucosaminidase) in the synovial fluid of the temporomandibular joints of TMD patients. It was revealead a positive correlation between these compounds detected and the degree of TMJ dysfunction as well as the amount of pain reported by the patients (Leonardi et al., 2011; Sato et al., 2003; Sato et al., 2003; Suzuki et al.,. 2002; Ogura et al.,. 2002; Tanaka et al., 2000; Appelgren et al., 1998; Kamada et al., 1993).

VII. Differential Diagnosis of TMDs

The differential diagnosis of temporo-mandibular disorders can be made with other common disorders that may mimic the symptoms of temporomandibular pain, which is specifically centred in and around the preauricular region, and dysfunction such as clicking or grating sounds associated with mandibular function and restricted mouth opening (Dimitroulis, 1998):

- dental pain such as tooth abscess, wisdom tooth eruption;
- disorders of the ears, nose, and sinuses such as mastoiditis, otitis;
- neuralgias such as herpes zoster neuralgia, trigeminal neuralgia;
- headaches; and
- diseases of the major salivary glands such as parotitis, sialadenitis

VIII. Treatment of TMDs (Surgical and Non-Surgical)

Management of TMDs often involves a multidisciplinary team, including dentists, orthodontists, psychologists, physical therapists, physicians and oral surgeons to address the condition of the patient. For most patients, the signs and symptoms of TMJ disorders improve over time and conservative treatment is considered to be the treatment of choice in 80% of the cases because the symptomatology of the condition often is improved by use of self-care and noninvasive treatments such as occlusal splints, acupuncture, intra-articular corticosteroid or anesthetic injection, pharmacologic treatment, psychological interventions such as hypnosis, cognitive behavior therapy, relaxation techniques or stress management. Although each treatment will be discussed separately, for optimal success they are best used in combination, depending on the patient's needs (De Boever et al., 2008; Ibric Cioranu et al., 1999).

VIII.1. Patient Education and Self Care

In aim to increase the compliance but also to reduce the symptom severity and the anxiety of the patients is extremely important to offer effective counseling by explaining the pain pathology and dysfunction and the co-factors involved (psychosocial and behavioral aspects, general diseases). The patient should also become aware of his responsibilities in the therapeutic process (compliance, motivation, coping) after discussing the management goals with him (De Boever et al., 2008).

VIII.2. Diet Change

Modifying the patient's diet by a nonchewing diet, such as liquid or pureed food, is performed in aim to reduce forces of mastication and TMJ loading (American Society of Temporomandibular Joint Surgeons, 2003).

VIII.3. Pharmacologic Intervention

It is common knowledge that the therapeutical approach of TMDs is directed to remove the cause of disorder, to eliminate symptomatology and to make an improvement to patient's life. The vast majority of patients are complaining by chronic pain, often caused by muscular component and rarely by the intercapsule, pain that is very stressful and extremely emotional (Cardelli et al., 2005).

Most current pharmacological management approaches in the treatment of orofacial pain conditions, including TMDs, involve the use of (Hersh et al., 2008; List et al., 2003):

- non-steroidal anti-inflammatory drugs (NSAIDs), salicylate and cyclo-oxygenase 2 (COX-2) inhibitors (e.g. Methyl salicylate, Rofecoxib 50 mg or Etoricoxib 120 mg once daily, Celecoxib 100 mg twice daily, diclofenac, naproxen, ibuprofen combined with other modalities, but not used alone) are useful for short-term (5 to 7 days) to reduce inflammation and treat pain episodes (such as following TMJ injury or surgery)
- corticosteroids,
- benzodiazepines (e.g. Clonazepam 0.25-1 mg daily; Prazepam 5 mg twice daily; Diazepam 2 mg daily) are used to reduce tension and spasm in the muscles affected by TMD
- nonbenzodiazepine sedative hypnotics,
- opioids (e.g. morphine, fentanyl, oxycodone, tramadol, hydrocodone, and methadone) are limited for long-term use as can lead to an increase in pain over time, as well as to dependence/addiction;
- skeletal muscle relaxants (e.g. cyclobenzaprine 10 mg daily usually for 2-3 weeks maximum)
- capsaicin (e.g. Capsaicin 0.025% four times a day),
- transdermal lidocaine (e.g. infiltration of 1 ml lidocaine 2% without vasoconstrictor in the involved muscle or in the trigger point),
- antidepressants (e.g. Amitriptyline 25 mg once daily, for 14 to 21 days) are used in low doses to help patients with TMD and are thought to be effective because they reduce muscle tension in patients who grind their teeth, and
- anticonvulsants (e.g. Diazepam 5 mg or Lorazepam 1 mg or Alprazolam 0.5 mg, once daily dosing for 5 to 10 days; Gabapentin 300 mg to 4200 mg daily).

A recent Cochrane systematic review has assessed the effectiveness of pharmacological interventions both alone and in combination with non-pharmacological therapy in relieving pain in patients with chronic TMD and concluded that there is insufficient evidence to support or not support the effectiveness of the reported drugs for the management of pain due to TMD (Mujakperuo et al., 2010).

Myofascial trigger-point injections with botulinum toxin type-A (Botox) has been shown to be an alternative method for controlling to reduce myofascial pain symptoms in bruxers (Guarda-Nardini et al., 2008; Arzul et al., 2012), while the intramuscular injections are able to reduce in 6 months the volume of the masticatory muscles (e.g. masseteric hypertrophy)(Monroy et al., 2006).

VIII.4. Occlusal Therapy

Nonoccluding splints, also called simple splints, such as night guards, bite guards typically made of a soft vinyl, primarily open the jaw, release muscle tension, and are widely used for bruxism control. Occluding splints, also called stabilization splints, yielding a more posterior positioning of the mandibular condyle during mouth closure, have also been used in aim to provide relaxation of the masticatory muscles, redistribute the forces applied to the jaws, protect the teeth and decrease the bruxism (Tanaka et al., 2004; Ekberg et al., 2003; Kreiner et al., 2001). However, no differences in terms of clinical outcome have been observed regarding stabilization splints in comparison with a soft splint, a non-occluding palatal splint, physical therapy, or body acupuncture (Türp et al., 2004). In patients with painful temporomandibular joint clicking, anterior positioning splint therapy appears to be the best treatment method for reduction of pain and joint sounds, compared with physical therapy, and physical therapy in addition to splint therapy (Madani and Mirmortazavi, 2011). An equivalent pain reduction with intra-articular injections with diclofenac sodium as compared with occlusal splints was found for the treatment of TMJ oteoarthritis (de Souza et al., 2012).

In 2005, in his review, the Cochrane Collaboration concluded that "there is insufficient evidence either for or against the use of stabilizatiaton splint therapy over other active interventions for the treatment of temporomandibular myofascial pain. However, it appears that stabilization splint therapy may be beneficial for reducing pain severity at rest and on palpation and depression when compared to no treatment."(Al-Ani et al., 2005). Similar results were reported by a more recent systematic review performed by Ebrahim et al. (2012) who revealed that splint therapy reduced pain in the TMJ area, with no significant differences in terms of quality of life or depression comparing with control groups receiving minimal or no treatment.

Adverse effects of occlusal splints have been reported, such as irreversible occlusal alterations, negative affection of the condyle-disk relation in patients who exhibit disk displacement with reduction and modification of breathing in patients with obstructive sleep apnea (Magdaleno & Ginestal, 2010).

Occlusal adjustment cannot be recommended for the management or prevention of TMD (Koh H, Robinson, 2003; Carlsson, 2010), but gross occlusal interferences (e.g. as a result of migration) may be removed to provide the patient with occlusal stability between the jaws after reduction of pain and normalization of the functional movements (De Boever et al, 2007).

VIII.5. Prosthetic Reconstructions

Prosthetic reconstruction should only be considered in the final phase of TMDs treatment after remission of the symptoms of pain and dysfunction as long treatment sessions may further stress the TMJ. If prosthetic treatment is needed due to particular cases, it should be kept as simple and as least invasive as possible (Boever et al, 2008).

VIII.6. Orthodontic Therapy

Psychosocial and aesthetic reasons are dominating factors indicating orthodontic treatment, as no scientific evidence exist that orthodontic treatment may cause or cure TMD (Michelotti and Iodice, 2010; Luther et al., 2010).

VIII.7. Physiotherapy

Physiotherapy is being widely used for patients with orofacial disorders, especially in the prevention or treatment of TMDs, hypomobility, or ankylosis, being of particular importance in the treatment of myofascial pain and internal derangement (Aggarwal & Keluskar, 2012). This adjunctive therapeutic modality has been introduced as a management of TMDs in aim to reduce patients'pain, reverse the TMJ dysfunction and restore optimal muscle and TMJ function, posture and activities of daily living to prevent recurrent episodes (De Boever et al., 2008).

There are three main groups of physiotherapy methods used in TMDs' management:

1) physical therapy (e.g. massage, therapeutic exercise, physical therapy, yoga)

- ✓ Treatment protocols including active exercises and manual mobilizations, alone or in combination, and associated often with a home exercise program, have been shown to be effective in the short term in increasing total vertical opening in people with TMDs (e.g. acute disk displacement, acute arthritis, or acute or chronic myofascial TMD)(Medlicott & Harris, 2006).
- ✓ Massage therapy for masseter and temporal muscles can be also a significant method of treatment of myofascial pain (Miernik et al., 2012).
- ✓ Beneficial results have been demonstrated by programs involving relaxation techniques and biofeedback, EMG training, proprioceptive reeducation in decreasing pain and increasing total vertical opening in people with acute or chronic myofascial or muscular in the short term and the long term (Medlicott & Harris, 2006).
- ✓ The Global Postural Re-education method, widely used in clinical practice, has reported benefits for rehabilitation of musculoskeletal dysfunctions, such as TMDs by reducing temporomandibular head and neck pain, and muscles electromyographic activity (Teodori et al., 2011; Maluf et al., 2010).
- ✓ In TMD resulting from acute disk displacement, acute arthritis, or acute myofascial TMD can be recommended combination programs (active exercises, manual therapy, postural correction, and relaxation techniques) in aim to decrease pain and impairment and increase total vertical opening in the short term (Medlicott & Harris, 2006).

2) electrotherapy methods (e.g. TENS: transcutaneous nerve stimulation, MENS: micro current electrical nerve stimulation, electro acupuncture, short wave diathermy, ultraviolet radiation, infrared radiation, ultrasound therapy, lasers).

- ✓ Due to the therapeutic properties of lasers (e.g. tissue repair, mediation of inflammatory processes, and analgesia in acute and chronic pain as well as improvement of local microcirculation), low-level laser can be efficiently used as an adjuvant therapeutic modality for the treatment of TMD, especially when the presence of inflammatory pain (Tengrungsun et al., 2012; Maia et al., 2012). As the synovial area is rather small in TMJ it was postulated that at least one single point is necessary to deliver an optimal dose of low-level laser therapy in this comparing with a minimum of three points of the synovial membranes of the knee and that must be irradiated to provide a sufficient dose (Bjordal et al., 2003).
- ✓ When the effectiveness of acupuncture for the symptomatic treatment of TMD was assesses, it was showed that classical acupuncture had a moderate positive influence beyond those of placebo, had positive effects similar to those of occlusal splint therapy and was more effective for TMD (Cho & Whang, 2010).

3) thermal modalities (e.g. ice massage, cold sprays) (Tokunaga et al., 1987)

VIII.8. Behavioural Therapy

Hypnosis and cognitive-behavioral interventions have been successfully used for pain management in cancer, osteoarthritis, sickle cell disease, fibromyalgia, non-cardiac chest pain (Elkins et al., 2012; Carlson et al., 2007), but also in temporomandibular disorders where the cognitive hypnotherapy should be part of the total treatment plan because of the role of psychosocial factors in the multifactorial etiology of TMDs (RDC/TMDs Axis II) (e.g. depression, somatization, stress, anxiety, neuroticism, pain-related impairment levels)(Aggarwal et al., 2011;Orlando et al., 2011). Moreover all the RDC/TMD axis II psychosocial scores have been inversely correlated with therapeutic outcome (weekly injections of hyaluronic acid immediately following arthrocentesis), being even more important than physical evaluation in terms of prognostic impact (Manfredini et al., 2013).

The principles underlying the cognitive behavioral therapy are based on the assumption that individual's thoughts are influencing his/her feelings and behaviors and that structured techniques are able to help individuals identifying and changing maladaptive thoughts/behaviors with development of teaching skills that individuals can then apply more when facing their problems (Calderon et al., 2011). It was also suggested that the combination of cognitive-behavioral therapy and amitriptyline, a tricyclic antidepressant, may be effective in reducing pain and depression levels as well as in improving the quality of life and sleep in patients with chronic TMDs (Calderon Pdos et al., 2011).

The dentist must be aware of the multidimensional approach to treatment because of the complexity of dysfunctions and should refer the patients for psychological evaluation and if indicated for therapy (De Boever et al., 2008).

VIII.9. Manual Reduction in Acute Disk Displacement

Acute displacement of the articular disk without reduction can be caused by trauma or general anesthesia. Due to articular disk that is leaving its physiologic position, acute disk displacement can lead to locked jaw in the open position and inhibition of swallowing and eating (Figure 5). Disk displacement should be reduced as soon as possible (Ibric Cioranu, 2002).

In aim to decrease pain and relax severe muscle spasm a local anesthetic or intravenous benzodiazepine is indicated. After the patient's head is stabilized (on a headrest or wall), manual reduction of the disk is performed by inserting the thumb into the patient's mouth, grasping under the chin, and simultaneously pushing down on the posterior teeth and pulling up on the chin. The mandibular condyle will be distracted downward, allowing the disk to move posteriorly into place (Buescher, 2007).

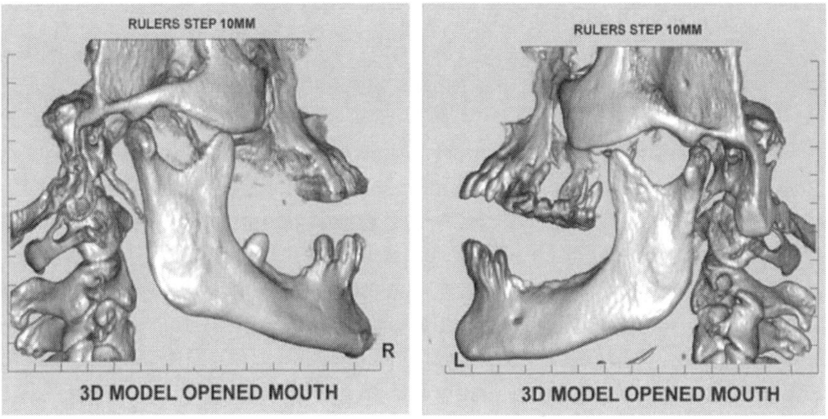

Figure 5. Computer tomography 3D reconstruction of a case with bilateral acute disk displacement.

VIII.10. Surgical Treatment

Surgery for treatment of TMDs has two main advantages: effectiveness and a rapid response. Patients considering surgery should be given the information necessary regarding the nature of their condition, potential surgical complications, side-effects, prognosis, and reasonable alternative treatment options (American Society of Temporomandibular Joint Surgeons, 2001).

a. Indications for TMJ Surgery

American Society of Temporomandibular Joint Surgeons (2001) suggests that surgical consultation should be offered immediately in cases of closed lock where delay in treatment can accelerate the progression of internal derangement/osteoarthritis, otherwise in patients in whom severe pain and dysfunction persists after 2-3 weeks of non-surgical therapy.

Accordig to Dimitroulis (2005) indications for surgery in the management of TMD may be divided into two groups:

1. Relative indications for cases where the option of non-surgical management plays a predominant role, such as TMJ internal derangement involving disc position and integrity and osteoarthrosis. Important clinical features for surgery include symptoms of pain and dysfunction (e.g. painful clicking, crepitus or hypomobility of the joint) well localised to the TMJ;
2. Absolute indications for TMJ surgery are uncommon disorders such as ankylosis, tumors, and growth anomalies.

The decision for surgical intervention must be based primarily on the clinical findings but also on other factors such as the impact of the disease on the well-being of the patient (e.g. patient's maximum interincisal mouth opening is insufficient for routine dental care), the prognosis of the disease when no treatment is provided, patient's psychosocial and cultural background. Additional investigations of the health status of bone (i.e., radiographs, tomograms or CT-scans) or soft tissue (i.e., MRI's) provides the essential data for an accurate diagnostic in considering, or not, the case for surgical intervention (Dimitroulis, 2005; Ibric Cioranu, 2002).

b. Minimally Invasive Procedures

The following minimally invasive surgical procedures have been accepted as effective methods for treatment of TMJ joints with internal derangement which presents as closed lock (i.e., disk displacement without reduction):

1. TMJ Arthrocentesis (i.e., acute onset closed lock)
2. TMJ Arthroscopy (i.e., long standing or chronic closed lock)

b.1. TMJ Arthrocentesis

Arthrocentesis, consists of washing the joint with the possibility of depositing a drug or other therapeutic substance (e.g. corticoids, hyaluronic acid) being considered a minimally invasive treatment modality, with low morbidity rates and highly effective in the treatment of patients with TMJ closed lock. The removal of chemical inflammatory mediators (interleukin (IL)-1beta, IL-6, IL-8, IL-11, and tumor necrosis factor (TNF)-alpha) and changes in intra-articular pressure are determinants of symptoms resolution (Gulen et al., 2009; Monje-Gil et al., 2012). Extensive amounts of joint effusion or erosive cortical changes of the condyle are generally associated with persistent joint pain after arthrocentesis (Honda et al., 2011).

The standard technique uses two punctures, in and out port, allowing low pressure irrigation of the superior joint space, but also a single puncture has been described (Manfredini et al., 2009; Shinohara et al., 2012). No significant differences in the treatment effectiveness for inflammatory-degenerative TMDs of a cycle of five weekly injections of arthrocentesis plus hyaluronic acid injections performed according to the classical two-needle or the single-needle were observed (Guarda-Nardini et al., 2012).

The conventional arthrocentesis under low pressure, with an elevated infusion bag or syringe, has been reported to be highly effective for treating acute closed lock without severe joint pain or bone change, but significantly lower in severe preoperative joint pain, chronic closed lock with osteoarthritis, synovitis, adhesions in the upper joint space, or bone changes (Sakamoto et al., 2000). A modified procedure with an arthrocentesis-like enforced manipulation technique followed by irrigation under high pressure (40 KPa) has been proposed by Yura et al. (2011).

An overview of the recent literature evaluating the effectiveness of arthrocentesis in the treatment of TMDs revealed the followings:

- Arthrocentesis provides long-term cure for degenerative TMJ (around 10 months period)(Onder et al., 2009);
- A treatment protocol consisting of cycle of 5 weekly arthrocenteses plus hyaluronic acid injections followed up for 1 year was more effective in patients with TMJ osteoarthritis older than 45 years with regard to almost all the considered symptoms (pain levels at rest and during chewing, subjective masticatory efficiency, functional limitation, perceived efficacy, jaw range of motion) in comparison with younger participants (Guarda-Nardini et al.,2012);
- In patients with early TMJ disc displacement without reduction, early treatment either with conservative methods (splint, heat, and exercise) or with arthrocentesis were similar with regard to the maximal mouth opening, lateral movements and protrusion. However, arthrocentesis seems to be superior regarding pain management being indicated especially in patients with increasing painful complaints despite other conservative treatments (Diraçoğlu et al., 2009);
- When the effectiveness of arthrocentesis was compared with arthroscopy no statistically significant difference was found between the interventions in terms of pain (both were associated with symptomatic improvement in 86% of patients), but a significant

difference in favour of arthroscopy was found in maximum incisal opening (Guo et al., 2009; Ahmed et al., 2012);

- Duration of locking and presence of preoperative degenerative changes are the most significant prognostic factors for successful arthrocentesis with and without sodium hyaluronate injection for the treatment of TMj disc displacement without reduction (Aktas et al., 2010);
- Intra-articular injections of long lasting opioids (buprenorphin) positively influence joint mobility and pain within the first week after arthrocentesis (Präger et al., 2007);

b.2. TMJ Arthroscopy

This highly technical intervention is used for simple procedures (e.g. simple arthroscopic lavage and lysis of adhesions within the TMJ) to complex operative procedures ranging from synovial sulcus ablation to disk suturing performed with miniaturised operative instruments. Both arthroscopic lavage and lysis and operative arthroscopy have been proved to be equally effective at decreasing pain in patients with TMJ internal derangement of any Wilkes stage. The best candidates for arthroscopy (had the highest decrease in pain and the highest increase in mouth opening among the stages) were patients classified as Wilkes stage IV presenting with chronic closed lock of the TMJ (González-García et al., 2011).

When different treatment options for permanent temporomandibular disc displacement were compared, no distinguishing effects on maximal mouth opening, pain or function impairment were found between arthroscopic surgery, arthrocentesis and physical therapy (Kropmans et al., 1999). Since then, Minakuchi et al., (2001) and Holmlund et al. (2001) found no difference between arthroscopic surgery and non-surgical treatments, or between arthroscopic surgery and discectomy in permanent temporomandibular joint disc displacement.

A recent systematic review of the Cochrane collaboration assessed the effectiveness of arthroscopy for the management of signs and symptoms in patients with TMDs. Both arthroscopy and nonsurgical treatments similarly reduced pain after 6 months. After 12 months arthroscopy was equal effective with arthrocentesis, but less effective than open surgery at reducing pain. After 12 months of postsurgical follow-up, arthroscopy led to greater improvement in maximum interincisal opening than arthrocentesis (Rigon et al., 2011). Similar results were reported also by Fridrich et al. (1996) and Goudot et al. (2000).

Several difficulties of surgical arthroscopy procedures should be mentioned the length of time of the procedure compared to conventional open joint interventions and the increased morbidity to intra-articular tissues which are constantly prodded and scraped by the numerous instruments including the arthroscopc itsclf (Dimitroulis, 2005).

c. Open Surgery

In the field of open TMJ surgery, the following procedures have been accepted as effective methods for treatment of TMDs (American Society of Temporomandibular Joint Surgeons, 2001): a) Condylotomy (indirect arthroplasty) (indicated in painful TMJ internal derangement where there is little or no restriction of mandibular function); b) Arthrotomy (advanced stages of TMJ internal derangement and osteoarthrosis) and c) Other procedures: Coronoidotomy/coronoidectomy, Styloidectomy (Eagle's Syndrome), Procedures for Recurrent Dislocation.

The severe internal derangements have been treated for many years with discectomy of the TMJ, but structural changes were observed in the condylar morphology subsequent to the intervention (Hansson et al., 1992). In the last half century, numerous attempts have been made to replace the missing disc with a barrier that acts as an original disc. Limited success was demonstrated in the 1980s by alloplastic implants such as Teflon-Proplast and in the 1990s byautogenous grafts such as temporalis muscle fascia, ear cartilage, and dermis (Dimitroulis, 2005). The abdominal dermis-fat graft was first introduced to TMJ surgery by Dimitroulis (2004) who used it as an interpositional graft in the management of TMJ ankylosis. However, presently is still concern that make the dermis-fat graft a less than ideal interpositional material for use in discectomized joint cavities and that there is still no ideal interpositional material that satisfies all the criteria for replacement of a missing articular disc following TMJ discectomy (Dimitroulis, 2011).

d. Postoperative Care

Essential for a good prognosis of TMDs treatment are a long-term follow-up and a good postoperative care that includes: wound care, thermal applications, non-chew diet regimen, medications, occlusal management, bruxism control, joint motion plans, active or passive joint exercise (American Society of Temporomandibular Joint Surgeons, 2001).

References

Aggarwal A, Keluskar V. Physiotherapy as an adjuvant therapy for treatment of TMJ disorders. *Gen Dent.* 2012;60:e119-22.

Aggarwal VR, Lovell K, Peters S, Javidi H, Joughin A, Goldthorpe J. Psychosocial interventions for the management of chronic orofacial pain. *Cochrane Database Syst Rev.* 2011:CD008456.

Ahmad M, Hollender L, Anderson Q, Kartha K, Ohrbach R, Truelove EL, John MT, Schiffman EL. Research diagnostic criteria for temporomandibular disorders (RDC/TMD): development of image analysis criteria and examiner reliability for image analysis. *Oral Surg Oral Med Oral Pathol Oral Radiol Endod.* 2009;107:844-60.

Ahn SJ, Kim TW, Lee DY, Nahm DS. Evaluation of internal derangement of the temporomandibular joint by panoramic radiographs compared with magnetic resonance imaging. *Am J Orthod Dentofacial Orthop.* 2006;129:479-85.

Aiken A, Bouloux G, Hudgins P. MR imaging of the temporomandibular joint. *Magn Reson Imaging Clin N Am.* 2012;20:397-412.

Akinbami BO. Evaluation of the mechanism and principles of management of temporomandibular joint dislocation review of literature and a proposed new classification of temporomandibular joint dislocation. *Head Face Med* 2011;7:10.

Aktas I, Yalcin S, Sencer S. Prognostic indicators of the outcome of arthrocentesis with and without sodium hyaluronate injection for the treatment of disc displacement without reduction: a magnetic resonance imaging study. *Int J Oral Maxillofac Surg.* 2010;39:1080-5.

Al-Ani Z, Gray RJ, Davies SJ, Sloan P, Glenny AM. Stabilization splint therapy for the treatment of temporomandibular myofascial pain: a systematic review. *J Dent Educ.* 2005;69:1242-50.

American Society of Temporomandibular Joint Surgeons. Guidelines for diagnosis and management of disorders involving the temporomandibular joint and related musculoskeletal structures. *Cranio.* 2003;21:68-76.

Appelgren A, Appelgren B, Kopp S, Lundeberg T, Theodorsson E. Substance P-associated increase of intra-articular temperature and pain threshold in the arthritic TMJ. *J Orofac Pain.* 1998;12:101-7.

Arzul L, Corre P, Khonsari RH, Mercier JM, Piot B. [Asymmetric hypertrophy of the masticatory muscles]. *Ann Chir Plast Esthet.* 2012;57:286-91.

Bagis B, Ayaz EA, Turgut S, Durkan R, Özcan M. Gender difference in prevalence of signs and symptoms of temporomandibular joint disorders:

a retrospective study on 243 consecutive patients. *Int J Med Sci.* 2012;9:539-44.

Bagis B, Ayaz EA, Turgut S, Durkan R, Özcan M. Gender difference in prevalence of signs and symptoms of temporomandibular joint disorders: a retrospective study on 243 consecutive patients. *Int J Med Sci.* 2012;9:539-44.

Balke Z, Rammelsberg P, Leckel M, Schmitter M. Prevalence of temporomandibular disorders: samples taken from attendees of medical health-care centers in the Islamic Republic of Iran. *J Orofac Pain.* 2010;24:361-6.

Barbosa Tde S, Miyakoda LS, Pocztaruk Rde L, Rocha CP, Gavião MB. Temporomandibular disorders and bruxism in childhood and adolescence: review of the literature. *Int J Pediatr Otorhinolaryngol.* 2008;72:299-314.

Barghan S, Merrill R, Tetradis S. Cone beam computed tomography imaging in the evaluation of the temporomandibular joint. *J Calif Dent Assoc.* 2010;38:33-9.

Barghan S, Tetradis S, Mallya S. Application of cone beam computed tomography for assessment of the temporomandibular joints. *Aust Dent J.* 2012;57:109-18.

Bas B, Yılmaz N, Gökce E, Akan H. Diagnostic value of ultrasonography in temporomandibular disorders. *J Oral Maxillofac Surg.* 2011;69:1304-10.

Bjordal JM, Couppé C, Chow RT, Tunér J, Ljunggren EA. A systematic review of low level laser therapy with location-specific doses for pain from chronic joint disorders. *Aust J Physiother.* 2003;49:107-16.

Bonjardim LR, Lopes-Filho RJ, Amado G, Albuquerque RL Jr, Goncalves SR. Association between symptoms of temporomandibular disorders and gender, morphological occlusion, and psychological factors in a group of university students. *Indian J Dent Res.* 2009;20:190-4.

Bouloux GF. Use of opioids in long-term management of temporomandibular joint dysfunction. *J Oral Maxillofac Surg.* 2011;69:1885-91.

Brandlmaier I, Bertram S, Rudisch A, Bodner G, Emshoff R. Temporomandibular joint osteoarthrosis diagnosed with high resolution ultrasonography versus magnetic resonance imaging: how reliable is high resolution ultrasonography? *J Oral Rehabil.* 2003;30:812-7.

Buescher JJ. Temporomandibular joint disorders. *Am Fam Physician.* 2007 Nov 15;76:1477-82.

Buescher JJ. Temporomandibular joint disorders. *Am Fam Physician.* 2007;76:1477-82.

Byahatti SM, Ramamurthy BR, Mubeen M, Agnihothri PG. Assessment of diagnostic accuracy of high-resolution ultrasonography in determination of temporomandibular joint internal derangement. *Indian J Dent Res.* 2010;21:189-94.

Calderon Pdos S, Tabaquim Mde L, Oliveira LC, Camargo AP, Ramos Netto Tde C, Cardelli P, Lattari M, Massaro P, Pollicita M, Barlattani A. *Pharmacologic treatment of the dysfunctional patient. Minerva Stomatol.* 2005;54:265-79.

Carlson CR. Psychological considerations for chronic orofacial pain. *Oral Maxillofac Surg Clin North Am.* 2008;20:185-95.

Carlsson GE. Epidemiology and treatment need for temporomandibular disorders. *J Orofac Pain.* 1999;13:232-7.

Carlsson GE. Some dogmas related to prosthodontics, temporomandibular disorders and occlusion. *Acta Odontol Scand.* 2010;68:313-22.

Casanova-Rosado JF, Medina-Solís CE, Vallejos-Sánchez AA, Casanova-Rosado AJ, Hernández-Prado B, Avila-Burgos L. Prevalence and associated factors for temporomandibular disorders in a group of Mexican adolescents and youth adults.*Clin Oral Investig.* 2006;10:42-9.

Cavalcanti RF, Studart LM, Kosminsky M, de Goes PS. Validation of the multimedia version of the RDC/TMD axis II questionnaire in Portuguese. *J Appl Oral Sci.* 2010;18:231-6.

Cho SH, Whang WW. Acupuncture for temporomandibular disorders: a systematic review. *J Orofac Pain.* 2010;24:152-62.

Choi YS, Choung PH, Moon HS, Kim SG. Temporomandibular disorders in 19-year-old Korean men. *J Oral Maxillofac Surg.* 2002;60:797-803.

Ciancaglini R, Gherlone EF, Radaelli G. Association between loss of occlusal support and symptoms of functional disturbances of the masticatory system. *J Oral Rehabil.* 1999;26:248-53.

Ciancaglini R, Gherlone EF, Radaelli G. Unilateral temporomandibular disorder and asymmetry of occlusal contacts. *J Prosthet Dent.* 2003;89:180-5.

Conti PC. Effectiveness of cognitive-behavioral therapy and amitriptyline in patients with chronic temporomandibular disorders: a pilot study. *Braz Dent J.* 2011;22:415-21.

Cooper BC, Kleinberg I. Examination of a large patient population for the presence of symptoms and signs of temporomandibular disorders. *Cranio.* 2007;25:114-26.

Coutinho A, Fenyo-Pereira M, Dib LL, Lima EN. The role of SPECT/CT with 99mTc-MDP image fusion to diagnose temporomandibular dysfunction. *Oral Surg Oral Med Oral Pathol Oral Radiol Endod.* 2006;101:224-30.

Dahlström L, Carlsson GE. Temporomandibular disorders and oral health-related quality of life. A systematic review. *Acta Odontol Scand.* 2010;68:80-5.

Dallanora AF, Grasel CE, Heine CP, Demarco FF, Pereira-Cenci T, Presta AA, Boscato N. Prevalence of temporomandibular disorders in a population of complete denture wearers. *Gerodontology.* 2012;29:e865-9.

de Andrade ED, Rizzatti-Barbosa CM, Pimenta Pinheiro ML. Pharmacological guidelines for managing temporomandibular disorders. *Braz J Oral Sci.* 2004;3:503-505.

De Boever JA, Nilner M, Orthlieb JD, Steenks MH; Educational Committee of the European Academy of Craniomandibular Disorders. Recommendations by the EACD for examination, diagnosis, and management of patients with temporomandibular disorders and orofacial pain by the general dental practitioner. *J Orofac Pain.* 2008;22:268-78.

de Leeuw R, editor. *Orofacial pain: guidelines for asssessment, diagnosis, and management. 4th ed.* Chicago: Quintessence Publishing;2008.

de Leeuw R, Eisenlohr-Moul T, Bertrand P. The association of smoking status with sleep disturbance, psychological functioning, and pain severity in patients with temporomandibular disorders. *J Orofac Pain.* 2013;27:32-41.

de Lucena LB, Kosminsky M, da Costa LJ, de Góes PS. Validation of the Portuguese version of the RDC/TMD Axis II questionnaire. *Braz Oral Res.* 2006;20:312-7.

de Oliveira AS, Dias EM, Contato RG, Berzin F. Prevalence study of signs and symptoms of temporomandibular disorder in Brazilian college students. *Braz Oral Res.* 2006 Jan-Mar;20:3-7.

de Souza RF, Lovato da Silva CH, Nasser M, Fedorowicz Z, Al-Muharraqi MA. Interventions for the management of temporomandibular joint osteoarthritis. *Cochrane Database Syst Rev.* 2012;4:CD007261.

Dijkstra PU, Kropmans TJ, Stegenga B. The association between generalized joint hypermobility and temporomandibular joint disorders: a systematic review. *J Dent Res.* 2002;81:158-63.

Dimitroulis G. A critical review of interpositional grafts following temporomandibular joint discectomy with an overview of the dermis-fat graft. *Int J Oral Maxillofac Surg.* 2011;40:561-8

Dimitroulis G. A new surgical classification for temporomandibular joint disorders. *Int J Oral Maxillofac Surg*. 2013;42:218-22.

Dimitroulis G. Temporomandibular disorders: a clinical update. *BMJ*. 1998;317:190-4.

Dimitroulis G. The interpositional dermis-fat graft in the management of temporomandibular joint ankylosis. *Int J Oral Maxillofac Surg*. 2004;33:755–60.

Dimitroulis G. The role of surgery in the management of disorders of the temporomandibular joint: a critical review of the literature. Part 2. *Int J Oral Maxillofac Surg*. 2005;34:231-7.

Diraçoğlu D, Saral IB, Keklik B, Kurt H, Emekli U, Ozçakar L, Karan A, Aksoy C. Arthrocentesis versus nonsurgical methods in the treatment of temporomandibular disc displacement without reduction. *Oral Surg Oral Med Oral Pathol Oral Radiol Endod.* 2009;108:3-8.

Dolwick MF. Clinical diagnosis of temporomandibular joint internal derangement and myofascial pain and dysfunction. *Oral Maxillofac Surg Clin North Am* 1989;1:1-6.

Dulcić N, Pandurić J, Kraljević S, Badel T, Celić R. Frequency of internal derangement of the temporomandibular joint in elderly individuals. *Eur J Med Res*. 2003;8:465-71.

Dulcić N, Pandurić J, Kraljević S, Badel T, Celić R. Incidence of temporomandibular disorders at tooth loss in the supporting zones. *Coll Antropol.*2003;27 Suppl 2:61-7.

Dworkin SF, LeResche L. Research diagmostic criteria for temporomandibular disorders: review, criteria, examinations and specifications, critique. *J Craniomandib Disord* 1992;6:301–55.

Dworkin SF, LeResche LR, DeRouen T, Von Korff M. Assessing clinical signs of temporomandibular disorders: reliability of clinical examiners. *J Prosthet Dent* 1990:63:574-79.

Dworkin SF, Sherman J, Mancl L, Ohrbach R, LeResche L, Truelove E. Reliability, validity, and clinical utility of the research diagnostic criteria for Temporomandibular Disorders Axis II Scales: depression, non-specific physical symptoms, and graded chronic pain. *J Orofac Pain*. 2002;16:207-20.

Ebrahim S, Montoya L, Busse JW, Carrasco-Labra A, Guyatt GH; Medically Unexplained Syndromes Research Group. The effectiveness of splint therapy in patients with temporomandibular disorders: a systematic review and meta-analysis. *J Am Dent Assoc*. 2012;143:847-57.

Egermark-Eriksson I, Ingervall B, Carlsson G E. The dependence of mandibular dysfunction in children on functional and morphologic malocclusion. *Am J Orthod* 1983;83: 187-194.

Ekberg E, Vallon D, Nilner M. The efficacy of appliance therapy in patients with temporomandibular disorders of mainly myogenous origin. A randomized, controlled, short-term trial. *J Orofac Pain*. 2003;17:133-9.

Eli I. Psychosocial factors in the etiology, diagnosis and management of temporomandibular disorders. *Alpha Omegan*. 2003;96:20-3.

Elkins G, Johnson A, Fisher W. Cognitive hypnotherapy for pain management. *Am J Clin Hypn*. 2012;54:294-310.

Emshoff R, Jank S, Bertram S, Rudisch A, Bodner G. Disk displacement of the temporomandibular joint: sonography versus MR imaging. *AJR Am J Roentgenol.* 2002;178:1557-62.

Ferrando M, Andreu Y, Galdón MJ, Durá E, Poveda R, Bagán JV. Psychological variables and temporomandibular disorders: distress, coping, and personality. *Oral Surg Oral Med Oral Pathol Oral Radiol Endod.* 2004;98:153-60.

Feteih RM. Signs and symptoms of temporomandibular disorders and oral parafunctions in urban Saudi Arabian adolescents: a research report. *Head Face Med.* 2006 Aug 16;2:25.

Fridrich KL, Wise JM, Zeitler DL. Prospective comparison of arthroscopy and arthrocentesis for temporomandibular joint disorders. *J Oral Maxillofac Surg.* 1996;54:816–820.

Gansky SA, Plesh O. Widespread pain and fibromyalgia in a biracial cohort study of young women. *J Rheumatol.* 2007;34:810–814.

Gesch D, Bernhardt O, Alte D, Schwahn C, Kocher T, John U, Hensel E. Prevalence of signs and symptoms of temporomandibular disorders in an urban and rural German population: results of a population-based Study of Health in Pomerania. *Quintessence Int*. 2004;35:143-50.

Gesch D, Bernhardt O, Kirbschus A. Association of malocclusion and functional occlusion with temporomandibular disorders (TMD) in adults: a systematic review of population-based studies. *Quintessence Int.* 2004;35:211-21.

Gonçalves DA, Dal Fabbro AL, Campos JA, Bigal ME, Speciali JG. Symptoms of temporomandibular disorders in the population: an epidemiological study. *J Orofac Pain*. 2010;24:270-8.

González-García R, Rodríguez-Campo FJ. Arthroscopic lysis and lavage versus operative arthroscopy in the outcome of temporomandibular joint internal derangement: a comparative study based on Wilkes stages. *J Oral Maxillofac Surg*. 2011;69:2513-24.

Goudot P, Jaquinet AR, Hugonnet S, Haefliger W, Richter M. Improvement of pain and function after arthroscopy and arthrocentesis of the temporomandibular joint: a comparative study. *J CranioMaxillofac Surg*. 2000;28:39–43.

Green CS, Laskin DM. Long term status of TMJ clicking in patients with myofascial pain dysfunction. *J Am Dent Assoc* 1988;117:461-5.

Guarda-Nardini L, Manfredini D, Salamone M, Salmaso L, Tonello S, Ferronato G. Efficacy of botulinum toxin in treating myofascial pain in bruxers: a controlled placebo pilot study. *Cranio*. 2008;26:126-35.

Guarda-Nardini L, Olivo M, Ferronato G, Salmaso L, Bonnini S, Manfredini D. Treatment effectiveness of arthrocentesis plus hyaluronic acid injections in different age groups of patients with temporomandibular joint osteoarthritis. *J Oral Maxillofac Surg*. 2012;70:2048-56.

Gulen H, Ataoglu H, Haliloglu S, Isik K. Proinflammatory cytokines in temporomandibular joint synovial fluid before and after arthrocentesis. *Oral Surg Oral Med Oral Pathol Oral Radiol Endod.* 2009;107:e1-4.

Guo C, Shi Z, Revington P. Arthrocentesis and lavage for treating temporomandibular joint disorders. *Cochrane Database Syst Rev*. 2009:CD004973.

Gurbuz O, Kursoglu P, Alatas G, Altinbas K. The prevalence of temporomandibular disorder signs in people with mental retardation. *J Oral Rehabil*. 2010;37:834-9.

Habets LLMH, Bezuur JN, van Ooij CP, Hansson TL. The orthopantomogram, an aid in diagnosis of temporomandibular joint problems. I. The factor of vertical magnification. *J Oral Rehabil* 1987;14:475-80.

Han DH, Park DY, Kim BI, Koh MY, Ahn YW, Kim JB. Global self-rating of oral health, concerns about oral health, and history of jaw injury related to temporomandibular joint symptoms in Korean adults. *J Orofac Pain*. 2011;25:308-16.

Hansson LG, Erksson L, Westesson P-L. Magnetic resonance evaluation after temporomandibular joint diskectomy. *Oral Surg Oral Med Oral Pathol*. 1992;74:801–10.

Hersh EV, Balasubramaniam R, Pinto A. Pharmacologic management of temporomandibular disorders. *Oral Maxillofac Surg Clin North Am*. 2008;20:197-210.

Honda K, Yasukawa Y, Fujiwara M, Abe T, Urade M. Causes of persistent joint pain after arthrocentesis of temporomandibular joint. *J Oral Maxillofac Surg*. 2011;69:2311-5.

Honey OB, Scarfe WC, Hilgers MJ, Klueber K, Silveira AM, Haskell BS, Farman AG. Accuracy of cone-beam computed tomography imaging of the temporomandibular joint: comparisons with panoramic radiology and linear tomography. *Am J Orthod Dentofacial Orthop* 2007;132:429-38.

Howard JA. Temporomandibular joint disorders in children. *Dent Clin North Am*. 2013;57:99-127.

Hussain AM, Packota G, Major PW, Flores-Mir C. Role of different imaging modalities in assessment of temporomandibular joint erosions and osteophytes: a systematic review. *Dentomaxillofac Radiol* 2008;37:63-71.

Ibric Cioranu V, Constantinescu MV, Slavescu DD, Podoleanu L. Tulburarile articulatiei temporo-mandibulare. *Diagnostic si tratament.* Editura Romcartexim, Bucuresti, 1999.

Ibric Cioranu V. *Chirurgia articulatiei temporo-mandibulare*. Editura Alma, Craiova, 2002.

Isong U, Gansky SA, Plesh O. Temporomandibular joint and muscle disorder-type pain in U.S. adults: the National Health Interview Survey. *J Orofac Pain*. 2008;22:317-22.

Jagger RG, Woolley SM, Savio L. Signs and symptoms of temporomandibular disorders in Ecuadorian Indians. *J Oral Rehabil.* 2004;31:293-7.

Kamada A, Fujita A, Kakudo K, Okazaki J, Ida M, Sakaki T. Changes in synovial fluid N-acetyl-beta-glucosaminidase activity in the human temporomandibular joint with dysfunction. *J Osaka Dent Univ.* 1993;27:107-11.

Karibe H, Goddard G, Aoyagi K, Kawakami T, Warita S, Shimazu K, Rudd PA, McNeill C. Comparison of subjective symptoms of temporomandibular disorders in young patients by age and gender. *Cranio*. 2012;30:114-20.

Katz J, Heft M. The epidemiology of self-reported TMJ sounds and pain in young adults in Israel. *J Public Health Dent.* 2002;62:177-9.

Koh H, Robinson PG. Occlusal adjustment for treating and preventing temporomandibular joint disorders. *Cochrane Database Syst Rev.* 2003:CD003812.

Koh KJ, List T, Petersson A, Rohlin M. Relationship between clinical and magnetic resonance imaging diagnoses and findings in degenerative and inflammatory temporomandibular joint diseases: a systematic literature review. *J Orofac Pain*. 2009;23:123-39.

Kreiner M, Betancor E, Clark GT. Occlusal stabilization appliances. Evidence of their efficacy. *J Am Dent Assoc*. 2001;132:770-7.

Kropmans JB, Dijkstra PU, Stegenga B, de Bont LG. Therapeutic outcome assessment in permanent temporomandibular joint disc displacement. *J Oral Rehabil*. 1999;26:357–363.

Kropmans TJ, Dijkstra PU, Stegenga B, de Bont LG. Therapeutic outcome assessment in permanent temporomandibular joint disc displacement. *J Oral Rehabil*. 1999;26:357-63.

Leonardi R, Almeida LE, Loreto C. Lubricin immunohistochemical expression in human temporomandibular joint disc with internal derangement. *J Oral Pathol Med*. 2011;40:587-92.

LeResche L, Mancl LA, Drangsholt MT, Huang G, Von Korff M. Predictors of onset of facial pain and temporomandibular disorders in early adolescence. *Pain*. 2007;129:269-78.

LeResche L. Epidemiology of temporomandibular disorders: implications for the investigation of etiologic factors. *Crit Rev Oral Biol Med.* 1997;8:291-305.

Li C, Su N, Yang X, Yang X, Shi Z, Li L. Ultrasonography for detection of disc displacement of temporomandibular joint: a systematic review and meta-analysis. *J Oral Maxillofac Surg*. 2012;70:1300-9.

Limchaichana N, Petersson A, Rohlin M. The efficacy of magnetic resonance imaging in the diagnosis of degenerative and inflammatory temporomandibular joint disorders: a systematic literature review. *Oral Surg Oral Med Oral Pathol Oral Radiol Endod*. 2006;102:521-36.

List T, Axelsson S, Leijon G. Pharmacologic interventions in the treatment of temporomandibular disorders, atypical facial pain, and burning mouth syndrome. A qualitative systematic review. *J Orofac Pain.* 2003;17:301-10.

Lobbezoo F, Visscher CM, Naeije M. Some remarks on the RDC/TMD Validation Project: report of an IADR/Toronto-2008 workshop discussion. *J Oral Rehabil.* 2010;37:779–83.

Look JO, John MT, Tai F, Huggins KH, Lenton PA, Truelove EL, Ohrbach R, Anderson GC, Shiffman EL. The Research Diagnostic Criteria For Temporomandibular Disorders. II: reliability of Axis I diagnoses and selected clinical measures. *J Orofac Pain.* 2010;24:25-34.

Luther F, Layton S, McDonald F. Orthodontics for treating emporomandibular joint (TMJ) disorders. *Cochrane Database Syst Rev.* 2010:CD006541.

Luther F. TMD and occlusion part I. Damned if we do? Occlusion: the interface of dentistry and orthodontics. *Br Dent J.* 2007a;202:E2; discussion 38-9.

Luther F. TMD and occlusion part II. Damned if we don't? Functional occlusal problems: TMD epidemiology in a wider context. *Br Dent J.* 2007b;202:E3;discussion 38-9.

Machado LP, Nery Cde G, Leles CR, Nery MB, Okeson JP. The prevalence of clinical diagnostic groups in patients with temporomandibular disorders. *Cranio.* 2009;27:194-9.

Mackie A, Lyons K. The role of occlusion in temporomandibular disorders—a review of the literature. *N Z Dent J.* 2008;104:54-9.

Madani AS, Mirmortazavi A. Comparison of three treatment options for painful temporomandibular joint clicking. *J Oral Sci.* 2011;53:349-54.

Magdaleno F, Ginestal E. Side effects of stabilization occlusal splints: a report of three cases and literature review. *Cranio.* 2010;28:128-35.

Magnusson T, Egermark I, Carlsson GE. A longitudinal epidemiologic study of signs and symptoms of temporomandibular disorders from 15 to 35 years of age. *J Orofac Pain.* 2000;14:310-9.

Magnusson T, Egermark I, Carlsson GE. Treatment received, treatment demand, and treatment need for temporomandibular disorders in 35-year-old subjects. *Cranio.* 2002;20:11-7.

Maia ML, Bonjardim LR, Quintans Jde S, Ribeiro MA, Maia LG, Conti PC. Effect of low-level laser therapy on pain levels in patients with temporomandibular disorders: a systematic review. *J Appl Oral Sci.* 2012;20:594-602.

Maluf SA, Moreno BG, Crivello O, Cabral CM, Bortolotti G, Marques AP. Global postural reeducation and static stretching exercises in the treatment of myogenic temporomandibular disorders: A randomized study. *J Manipulative Physiol Ther.* 2010;33:500-7.

Manfredini D, Arveda N, Guarda-Nardini L, Segù M, Collesano V. Distribution of diagnoses in a population of patients with temporomandibular disorders. *Oral Surg Oral Med Oral Pathol Oral Radiol.* 2012;114:e35-41.

Manfredini D, Favero L, Del Giudice A, Masiero S, Stellini E, Guarda-Nardini L. Axis II psychosocial findings predict effectiveness of TMJ hyaluronic acid injections. *Int J Oral Maxillofac Surg.* 2013;42:364-8.

Manfredini D, Guarda-Nardini L, Ferronato G. Single-needle temporomandibular joint arthrocentesis with hyaluronic acid injections.

Preliminary data after a five-injection protocol. *Minerva Stomatol.* 2009;58:471-8.

Manfredini D, Guarda-Nardini L. Ultrasonography of the temporomandibular joint: a literature review. *Int J Oral Maxillofac Surg.* 2009;38:1229-36.

Manfredini D, Landi N, Bandettini Di Poggio A, Dell'Osso L, Bosco M. A critical review on the importance of psychological factors in temporomandibular disorders. *Minerva Stomatol.* 2003;52:321-30.

Manfredini D, Lobbezoo F. Relationship between bruxism and temporomandibular disorders: a systematic review of literature from 1998 to 2008. *Oral Surg Oral Med Oral Pathol Oral Radiol Endod.* 2010;109:e26-50.

Manfredini D, Segù M, Bertacci A, Binotti G, Bosco M. Diagnosis of temporomandibular disorders according to RDC/TMD axis I findings, a multicenter Italian study. *Minerva Stomatol.* 2004;53:429-38.

Marklund S, Wänman A. Incidence and prevalence of temporomandibular joint pain and dysfunction. A one-year prospective study of university students. *Acta Odontol Scand.* 2007;65:119-27.

Medlicott MS, Harris SR. A systematic review of the effectiveness of exercise, manual therapy, electrotherapy, relaxation training, and biofeedback in the management of temporomandibular disorder. *Phys Ther.* 2006;86:955-73.

Melis M, Secci S, Ceneviz C. Use of ultrasonography for the diagnosis of temporomandibular joint disorders: a review. *Am J Dent.* 2007;20:73-8.

Meloto CB, Serrano PO, Ribeiro-DaSilva MC, Rizzatti-Barbosa CM. Genomics and the new perspectives for temporomandibular disorders. *Arch Oral Biol.* 2011;56:1181-91.

Michelotti A, Iodice G. The role of orthodontics in temporomandibular disorders. *J Oral Rehabil.* 2010;37:411-29.

Minakuchi H, Kuboki T, Matsuka Y, Maekawa K, Yatani H, Yamashita A. Randomized controlled evaluation of non-surgical treatments for temporomandibular joint anterior disk displacement without reduction. *J Dent Res.* 2001;80:924-8.

Miyake R, Ohkubo R, Takehara J, Morita M. Oral parafunctions and association with symptoms of temporomandibular disorders in Japanese university students. *J Oral Rehabil.* 2004;31:518-23.

Mohlin B, Axelsson S, Paulin G, Pietilä T, Bondemark L, Brattström V, Hansen K, Holm AK. TMD in relation to malocclusion and orthodontic treatment. *Angle Orthod.* 2007;77:542-8.

Mohlin B, Thilander B. The importance of the relationship between malocclusion and mandibular dysfunction and some clinical applications in adults. *Eur J Orthod* 1984;6: 192-204.

Monje-Gil F, Nitzan D, González-Garcia R. Temporomandibular joint arthrocentesis. Review of the literature. *Med Oral Patol Oral Cir Bucal.* 2012;17:e575-81.

Monroy PG, da Fonseca MA. The use of botulinum toxin-a in the treatment of severe bruxism in a patient with autism: a case report. *Spec Care Dentist.* 2006;26:37-9.

Moyaho-Bernal A, Lara-Muñoz Mdel C, Espinosa-De Santillana I, Etchegoyen G. Prevalence of signs and symptoms of temporomandibular disorders in children in the State of Puebla, Mexico, evaluated with the research diagnostic criteria for temporomandibular disorders (RDC/TMD). *Acta Odontol Latinoam.* 2010;23:228-33.

Muhtaroğullari M, Demirel F, Saygili G. Temporomandibular disorders in Turkish children with mixed and primary dentition: prevalence of signs and symptoms. *Turk J Pediatr.* 2004;46:159-63.

Mujakperuo HR, Watson M, Morrison R, Macfarlane TV. Pharmacological interventions for pain in patients with temporomandibular disorders. *Cochrane Database Syst Rev.* 2010:CD004715.

Nassif NJ, Al-Salleeh F, Al-Admawi M. The prevalence and treatment needs of symptoms and signs of temporomandibular disorders among young adult males. *J OralRehabil.* 2003;30:944-50.

Nekora-Azak A, Evlioglu G, Ordulu M, Işsever H. Prevalence of symptoms associated with temporomandibular disorders in a Turkish population. *J Oral Rehabil.* 2006;33:81-4.

Nilsson IM, Drangsholt M, List T. Impact of temporomandibular disorder pain in adolescents: differences by age and gender. *J Orofac Pain.* 2009;23:115-22.

Nilsson IM, List T, Drangsholt M. Prevalence of temporomandibular pain and subsequent dental treatment in Swedish adolescents. *J Orofac Pain.* 2005;19:144-50.

Nomura K, Vitti M, Oliveira AS, Chaves TC, Semprini M, Siéssere S, Hallak JE, Regalo SC. Use of the Fonseca's questionnaire to assess the prevalence and severity of temporomandibular disorders in Brazilian dental undergraduates. *Braz Dent J.* 2007;18:163-7.

Oakley M, Vieira AR. The many faces of the genetics contribution to temporomandibular joint disorder. *Orthod Craniofac Res.* 2008;11: 125-35.

Ogura N, Tobe M, Sakamaki H, Kujiraoka H, Akiba M, Abiko Y, Nagura H. Interleukin-1 beta induces interleukin-6 mRNA expression and protein production in synovial cells from human temporomandibular joint. *J Oral Pathol Med.* 2002;31:353-60.

Ogura T, Morinushi T, Ohno H, Sumi K, Hatada K. An epidemiological study of TMJ dysfunction syndrome in adolescents. *J Pedod.* 1985;10:22-35.

Ohrbach R, Gale EN. Pressure pain thresholds, clinical assessment and differential diagnosis: reliability and validity in patients with myogenic pain. *Pain* 1989;39:157-69.

Ohrbach R, Turner JA, Sherman JJ, Mancl LA, Truelove EL, Schiffman EL, Dworkin SF. The Research Diagnostic Criteria for Temporomandibular Disorders. IV: evaluation of psychometric properties of the Axis II measures. *J Orofac Pain.* 2010;24:48-62.

Onder ME, Tüz HH, Koçyiğit D, Kişnişci RS. Long-term results of arthrocentesis in degenerative temporomandibular disorders. *Oral Surg Oral Med Oral Pathol Oral Radiol Endod.* 2009;107:e1-5.

Orlando B, Manfredini D, Salvetti G, Bosco M. Evaluation of the effectiveness of biobehavioral therapy in the treatment of temporomandibular disorders: a literature review. *Behav Med.* 2007;33:101-18.

Otuyemi OD, Owotade FJ, Ugboko VI, Ndukwe KC, Olusile OA. Prevalence of signs and symptoms of temporomandibular disorders in young Nigerian adults. *J Orthod.* 2000;27:61-5.

Plesh O, Adams SH, Gansky SA. Racial/Ethnic and gender prevalences in reported common pains in a national sample. *J Orofac Pain.* 2011; 25:25-31.

Plesh O, Crawford PB, Gansky SA. Chronic pain in a biracial population of young women. *Pain.* 2002;99:515–523.

Poveda Roda R, Bagan JV, Díaz Fernández JM, Hernández Bazán S, Jiménez Soriano Y. Review of temporomandibular joint pathology. Part I: classification, epidemiology and risk factors. *Med Oral Patol Oral Cir Bucal.* 2007;12:E292-8.

Poveda-Roda R, Bagán JV, Jiménez-Soriano Y, Fons-Font A. Retrospective study of a series of 850 patients with temporomandibular dysfunction (TMD). Clinical and radiological findings. *Med Oral Patol Oral Cir Bucal.* 2009;14:e628-34.

Poveda-Roda R, Bagan JV, Sanchis JM, Carbonell E. Temporomandibular disorders. A case-control study. *Med Oral Patol Oral Cir Bucal.* 2012 Sep 1;17:e794-800.

Pow EH, Leung KC, McMillan AS. Prevalence of symptoms associated with temporomandibular disorders in Hong Kong Chinese. *J Orofac Pain.* 2001;15:228-34.

Ribeiro-Rotta RF, Marques KD, Pacheco MJ, Leles CR. Do computed tomography and magnetic resonance imaging add to temporomandibular joint disorder treatment? A systematic review of diagnostic efficacy. *J Oral Rehabil.* 2011;38:120-35.

Rigon M, Pereira LM, Bortoluzzi MC, Loguercio AD, Ramos AL, Cardoso JR. Arthroscopy for temporomandibular disorders. *Cochrane Database Syst Rev.* 2011:CD006385.

Riley JL, Gilbert GH, Heft MW. Orofacial pain:recial and sex differences among older adults. *J Publ Health Dent.* 2002;62:132–139.

Ruf S, Pancherz H. Is orthopantomography reliable for TMJ diagnosis? An experimental study on a dry skull. *J Orofac Pain*1995;9:365-74.

Rutkiewicz T, Könönen M, Suominen-Taipale L, Nordblad A, Alanen P. Occurrence of clinical signs of temporomandibular disorders in adult Finns. *J Orofac Pain.* 2006;20:208-17.

Sakamoto I, Yoda T, Tsukahara H, Imai H, Enomoto S. Comparison of the effectiveness of arthrocentesis in acute and chronic closed lock: analysis of clinical and arthroscopic findings. *Cranio.* 2000;18:264-71.

Sanders AE, Maixner W, Nackley AG, Diatchenko L, By K, Miller VE, Slade GD. Excess risk of temporomandibular disorder associated with cigarette smoking in young adults. *J Pain.* 2012;13:21-31.

Sanders AE, Slade GD. Gender modifies effect of perceived stress on orofacial pain symptoms: National Survey of Adult Oral Health. *J Orofac Pain.* 2011;25:317-26.

Saridin CP, Raijmakers PG, Al Shamma S, Tuinzing DB, Becking AG. Comparison of different analytical methods used for analyzing SPECT scans of patients with unilateral condylar hyperactivity. *Int J Oral Maxillofac Surg.* 2009;38:942-6.

Saridin CP, Raijmakers PG, Tuinzing DB, Becking AG. Comparison of planar bone scintigraphy and single photon emission computed tomography in patients suspected of having unilateral condylar hyperactivity. *Oral Surg Oral Med Oral Pathol Oral Radiol Endod.* 2008;106:426-32.

Sarikaya I, Sarikaya A, Holder LE. The role of single photon emission computed tomography in bone imaging. *Semin Nucl Med.* 2001;31:3-16.

Sato J, Segami N, Nishimura M, Demura N, Yoshimura H, Yoshitake Y, Nishikawa K. Expression of interleukin 6 in synovial tissues in patients with internal derangement of the temporomandibular joint. *Br J Oral Maxillofac Surg*. 2003;41:95-101.

Sato J, Segami N, Nishimura M, Kaneyama K, Demura N, Yoshimura H. Relation between the expression of vascular endothelial growth factor in synovial tissues and the extent of joint effusion seen on magnetic resonance imaging in patients with internal derangement of the temporomandibular joint. *Br J Oral Maxillofac Surg*. 2003;41:88-94.

Schiffman EL, Ohrbach R, Truelove EL, Tai F, Anderson GC, Pan W, Gonzalez YM, John MT, Sommers E, List T, Velly AM, Kang W, Look JO. The Research Diagnostic Criteria for Temporomandibular Disorders. V: methods used to establish and validate revised Axis I diagnostic algorithms. *J Orofac Pain*. 2010;24:63-78.

Schmid-Schwap M, Bristela M, Kundi M, Piehslinger E. Sex-specific differences in patients with temporomandibular disorders. *J Orofac Pain*. 2013;27:42-50.

Seedorf H, Seetzen F, Scholz A, Sadat-Khonsari MR, Kirsch I, Jüde HD. Impactf posterior occlusal support on the condylar position. *J Oral Rehabil*. 2004;31:759-63

Shinohara EH, Pardo-Kaba SC, Martini MZ, Horikawa FK. Single puncture for TMJ arthrocentesis: An effective technique for hydraulic distention of the superior joint space. *Natl J Maxillofac Surg*. 2012;3:96-7.

Slade GD, Diatchenko L, Bhalang K, Sigurdsson A, Fillingim RB, Belfer I, Max MB, Goldman D, Maixner W. Influence of psychological factors on risk of temporomandibular disorders. *J Dent Res*. 2007;86:1120-5.

Sönmez H, Sari S, Oksak Oray G, Camdeviren H. Prevalence of temporomandibular dysfunction in Turkish children with mixed and permanent dentition. *J Oral Rehabil*. 2001;28:280-5.

Storm C, Wänman A. Temporomandibular disorders, headaches, and cervical pain among females in a Sami population. *Acta Odontol Scand*. 2006;64:319-25.

Suvinen TI, Reade PC, Kemppainen P, Könönen M, Dworkin SF. Review of aetiological concepts of temporomandibular pain disorders: towards a biopsychosocial model for integration of physical disorder factors with psychological and psychosocial illness impact factors. *Eur J Pain*. 2005;9:613-33.

Suzuki T, Segami N, Nishimura M, Nojima T. Co-expression of interleukin-1beta and tumor necrosis factor alpha in synovial tissues and synovial fluids of temporomandibular joint with internal derangement: comparison with histological grading of synovial inflammation. *J Oral Pathol Med.* 2002;31:549-57.

Takayama Y, Miura E, Yuasa M, Kobayashi K, Hosoi T. Comparison of occlusal condition and prevalence of bone change in the condyle of patients with and without temporomandibular disorders. *Oral Surg Oral Med Oral Pathol Oral Radiol Endod.* 2008;105:104-12.

Tanaka A, Kawashiri S, Kumagai S, Takatsuka S, Narinobou M, Nakagawa K, Tanaka S. Expression of matrix metalloproteinase-2 in osteoarthritic fibrocartilage from human mandibular condyle. *J Oral Pathol Med.* 2000;29:314-20.

Tanaka EE, Arita ES, Shibayama B. Occlusal stabilization appliance: evaluation of its efficacy in the treatment of temporomandibular disorders. *J Appl Oral Sci.* 2004;12:238-43.

Tengrungsun T, Mitriattanakul S, Buranaprasertsuk P, Suddhasthir T. Is low level laser effective for the treatment of orofacial pain?: A systematic review. *Cranio.* 2012;30:280-5.

Teodori RM, Negri JR, Cruz MC, Marques AP. Global Postural Re-education: a literature review. *Rev Bras Fisioter.* 2011;15:185-9.

Thilander B, Bjerklin K. Posterior crossbite and temporomandibular disorders (TMDs): need for orthodontic treatment? *Eur J Orthod.* 2012;34:667-73.

Tognini F, Manfredini D, Melchiorre D, Bosco M. Comparison of ultrasonography and magnetic resonance imaging in the evaluation of temporomandibular joint disc displacement. *J Oral Rehabil.* 2005;32:248-53.

Tognini F, Manfredini D, Melchiorre D, Zampa V, Bosco M. Ultrasonographic vs magnetic resonance imaging findings of temporomandibular joint effusion. *Minerva Stomatol.* 2003;52:365-70, 370-2.

Tokunaga T, Yuba N, Ohta K, Kiyohara S, Tanaka M, Kawazoe T. [The effect of thermal therapies on the masseter muscle for temporomandibular joint disorders]. *Nihon Hotetsu Shika Gakkai Zasshi.* 1987;31:770-7.

Tomas X, Pomes J, Berenguer J, Quinto L, Nicolau C, Mercader JM, Castro V. MR imaging of temporomandibular joint dysfunction: a pictorial review. *Radiographics.* 2006;26:765-81.

Tuerlings V, Limme M. The prevalence of temporomandibular joint dysfunction in the mixed dentition. *Eur J Orthod.* 2004;26:311-20.

Türp JC, Komine F, Hugger A. Efficacy of stabilization splints for the management of patients with masticatory muscle pain: a qualitative systematic review. *Clin Oral Investig.* 2004;8:179-95.

Türp JC, Schindler H. The dental occlusion as a suspected cause for TMDs: epidemiological and etiological considerations. *J Oral Rehabil.* 2012;39:502-12.

Tvrdy P. Methods of imaging in the diagnosis of temporomandibular joint disorders. *Biomed Pap Med Fac Univ Palacky Olomouc Czech Repub.* 2007;151:133-6.

Uhac I, Kovac Z, Muhvić-Urek M, Kovacević D, Francisković T, Simunović-Soskić M. The prevalence of temporomandibular disorders in war veterans with post-traumatic stress disorder. *Mil Med.* 2006;171:1147-9.

Unell L, Johansson A, Carlsson GE, Halling A, Söderfeldt B. Changes in reported orofacial symptoms over a ten-year period as reflected in two cohorts of fifty-year-old subjects. *Acta Odontol Scand.* 2006;64:202-8.

Velly AM, Gornitsky M, Philippe P. Contributing factors to chronic myofascial pain: a case-control study. *Pain.* 2003;104:491-9.

Villanueva-Alcojol L, Monje F, González-García R. Hyperplasia of the mandibular condyle: clinical, histopathologic, and treatment considerations in a series of 36 patients. *J Oral Maxillofac Surg.* 2011;69:447-55.

Weingarten TN, Iverson BC, Shi Y, Schroeder DR, Warner DO, Reid KI. Impact of tobacco use on the symptoms of painful temporomandibular joint disorders. *Pain.* 2009;147:67-71.

Westesson PL, Katzberg RW, Tallents RH, Sanchez-Woodworth RE, Svensson SA. CT and MR of the temporomandibular joint: comparison with autopsy specimens. *AJR Am J Roentgenol* 1987;148:1165-71.

Wilkes CH. Internal derangements of the temporomandibular joint: pathological variations. *Arch Otolaryngol Head Neck Surg* 1989;115:469–77.

Wu N, Hirsch C. Temporomandibular disorders in German and Chinese adolescents. *J Orofac Orthop.* 2010;71:187-98.

Yap AU, Dworkin SF, Chua EK, List T, Tan KB, Tan HH. Prevalence of temporomandibular disorder subtypes, psychologic distress, and psychosocial dysfunction in Asian patients. *J Orofac Pain.* 2003;17:21-8.

Yura S, Ooi K, Izumiyama Y. Relationship between the Effectiveness of Arthrocentesis under Sufficient Pressure and Conditions of the Temporomandibular Joint. *ISRN Dent.* 2011;2011:376475.

In: Craniofacial Disorders
Editors: Sh.B. Turner and J. E. Miller
ISBN: 978-1-62618-923-2

Chapter 2

Paradigm Shift in Pre-Surgical Orthopedics for Infants with Oral Clefts

Supakit Peanchitlertkajorn*
Private Practice, Hayward Braces
Cleft Orthodontist, Kaiser Permanente, CA, US

Abstract

Pre-surgical orthopedic treatments have been employed in infants born with oral clefts for over half a century. There are a variety of technics described in the literature. The early treatment goals included feeding assistance as well as aligning alveolar segments. Advocates of the treatment claimed that long-term benefits included speech and orthodontic benefits. However, recent studies have shown that such long-term benefits could not be substantiated. The ability to mold neonatal cartilages has given rise to a new pre-surgical technic called Nasoalveolar Molding (NAM). The NAM technic added nasal cartilage molding as part of its treatment goal. This has allowed surgeons to consistently obtain improved esthetics in the nasolabial area. The popularity of NAM in recent years has shifted traditional treatment goals of the pre-surgical orthopedic treatments to focus more on enhancing nasolabial esthetics. In this chapter, we will discuss publications that deal with pre-surgical

* Corresponding author: email: supakit@att.net.

orthopedics for infants with oral cleft, particularly those dealing with NAM treatment.

Introduction

Presurgical orthopedic treatment (PSOT)for infants with oral cleft remains a controversial subject since its introduction by McNeil in 1950 [1]. McNeil reported a method of repositioning the displaced maxillary segments to reduce the size of the initial alveolar cleft and approximate the lip segments. Since then, a variety of PSOT technics have been described in the literature. Hotz and Gnoinski advocated a passive form of an orthopedic device [2]. They fabricated a palatal plate that covered the alveolar segments passively. The plate was periodically adjusted so that the alveolar segments could be moved to reduce the cleft width. They also recommended wearing the appliance post-surgically to prevent a collapse of alveolar segments that could lead to a dental crossbite. The appliance also covered the soft palate area in hope of a better speech development. Furthermore, they believed that wearing the passive orthopedic plate could normalize the tongue position and prevent the tongue intrusion into the cleft area. In turn, the plate could reduce the degree of the curvature of the maxillary segments toward the nasal cavity. In 1980, Latham described an active form of an oral device with a screw designed to approximate all maxillary segments [3]. This technic utilized a heavy force to achieve the alveolar movement. The outcome could be achieved in a shorter period of time compared to other technics. This PSOT technic was usually performed in combination with lip adhesion and gingivoperiostoplasty [4]. However, long-term outcome studies showed a severe maxillary hypoplasia in patients who were treated by this combined approach [5-8].

The proponents of these technics have claimed several benefits including maxillary growth, feeding, and speech. These reports on the benefits of PSOT were largely anecdotal. In addition, proper evaluation of these claims has been difficult due to the poor design of early studies. This chapter will explore recent studies with more scientifically rigorous designs that evaluated these claims. Although the principles of PSOT for infants with oral clefts began with the traditional paradigm of reducing the cleft width to facilitate lip surgery, most early studies focused on maxillary arch form and maxillary growth rather than nasolabial esthetic outcome. This chapter will present currently available evidence in the literature regarding the effects of PSOT on nasolabial esthetics.

Current Use of PSOT

An international survey conducted in 2008 revealed that 71% of respondents used PSOT occasionally. However, there was no consensus regarding the type of orthopedic device used. According to this survey, the passive form of orthopedic devices (Hotz plate) was used more frequently than the active form (Latham appliance). The survey also reported that 38% of the respondents performed pre-surigcal nasoalveolar molding (NAM) on some of their patients especially those with a more severe condition [9].

PSOT Effects on Maxillary Growth and Arch Form

The advocates of PSOT have long claimed that PSOT has a favorable effect on maxillary growth in patients with oral clefts. Ross attempted to verify this claim by studying an international sample of 97 patients with complete unilateral cleft lip and palate (CUCLP) [10]. All patients were treated with either active or passive PSOT. He compared this sample to a control group that did not receive any PSOT. The author found no long-term facial growth benefits from PSOT. He further reported that the facial growth difference between the active PSOT group and the passive PSOT group was rather negligible.

Eurocleft is a series of retrospective studies comparing treatment outcomes from 6 European centers, each using a different treatment protocol [11]. The studies used dental plaster models and cephalograms to compare maxillary growth and arch form. While it is impossible to establish a cause and effect relationship between a certain protocol feature and the outcome due to treatment variables in the protocol, the comparison of outcomes in these studies provided a basis for understanding certain protocol features that might be associated with favorable or unfavorable outcomes. The PSOT technics performed in the participating centers included Hotz plate, extraoral strapping, and extraoral traction. If we assumed that PSOT had any positive treatment effect, then treatment outcomes from the protocols with PSOT should be ranked rather high in these comparisons. However, the findings from these studies showed that the protocol using Hotz plate was ranked in the middle while the protocol with extraoral strapping and traction was ranked quite low. The authors concluded that PSOT has no effect (neither positive nor negative) on maxillary growth and arch form [12-14]. These researchers also longitudinally followed the same group of patients up to 17 years of age and performed another comparison of treatment outcomes [15]. They reported the findings to be very similar to their original studies [16-17]. The success of the Eurocleft studies had led several American cleft centers to conduct similar studies of their own. There were 5 participating centers with a sample size of 30-40 patients. Dental study models and cephalograms were also used for comparisons. 3 centers performed PSOT. One center used Modified McNeil plate only before lip surgery. Another center also used a Modified McNeil plate both before and after lip surgery. One center employed a molding plate and used it in combination with primary alveolar bone grafting. The results from these studies also suggested PSOT does not have any long-term effect on maxillary growth and arch form [18-20].

Adali and colleagues evaluated the effect of PSOT on maxillary arch form up to 6 months of age [21]. 75 consecutive infants with CUCLP were recruited to take part in this study. 14 patients received PSOT whereas the remainder did not. They concluded that PSOT does not appear to have any effect on arch form. They also noted that lip repair produced greater effect on arch form than did PSOT.

20 Japanese infants with CUCLP participated in studies evaluating the effect of the Hotz plate. 12 randomly selected infants wore Hotz plates for 18 months while 8 other infants did not. They found that the group that received PSOT had a more normal palatal curvature when compared to the group without PSOT. In addition, they reported that the width of the palate was larger in the group with Hotz plate. However, they found no difference between the anteroposterior distances of the palate in the two groups. Similarly, the alveolar arch form did not differ between the two groups. These results were similar for both 18 months and 4 years of age [22-23].

One of the most scientifically rigorous studies on the effects of PSOT available in the literature was published in a series of articles known as Dutchcleft studies. The study was designed as a randomized clinical trial. 54 infants with CUCLP participated in this study. The treatment group (27 infants) received Hotz plate at birth and continuing for 12 months of their life before surgical correction, then for several months following. The remainder did not receive any form of PSOT before the surgical repair. The studies followed the patients for about 6 years. Dental models and cephalograms were used to evaluate the maxillary growth and arch form. The authors found only a temporary effect on maxillary arch dimensions. They found no observable benefit on maxillary growth in primary dentition. Additionally, the use of PSOT does not appear to prevent the collapse of alveolar segments [25-28].

Based on these recent studies, it appears that PSOT does not have any effect on sagittal growth of the maxilla. Though PSOT might reduce the degree of palatal curvature, it does not improve the maxillary arch form.

PSOT and Speech

The early reports regarding the effects of PSOT on speech development have been mostly anecdotal. Recent studies however employed more scientifically sound methods to substantiate the speech benefits of the PSOT. Hardin-Jones and colleagues investigated the effects of a palatal obturator[29]. In this study, 14 infants with cleft palate only wore palatal obturators prior to palatal surgery while the other 14 infants did not. Their findings did not support the speech benefit of this type of obturator.

The Dutchcleft study is the first randomized clinical trial evaluating the effects of PSOT on different aspects of speech development. The findings from these studies revealed a more favorable speech development in patients with PSOT. They received a higher speech intelligibility rating at age 2.5 years [30]. They also showed more normal phonological development between age 2 and 3 years [31]. They produced longer sentences (better expressive language skill) than non-PSOT children at the ages of 2.5 and 3 years. However, At 6 years of age, both groups exhibited similar expressive language skills [32].

It should be noted that the appliance used in the Dutchcleft study covered both hard and soft palate whereas the appliance used in the previous study only covered the hard palate. The difference in appliance design may contribute to the contrast in findings. This implies that the type and design of PSOT may be an important variable on treatment effects.

PSOT and Feeding

The claim that PSOT could facilitate feeding in infants with oral cleftsis one of the most common assertions with very little supporting evidences. The Dutchcleft studies also investigated the effect of PSOT on feeding [33]. They reported no difference in feeding velocity between the group of infants who received PSOT and the group without PSOT. In addition, the PSOT group gained weight quicker initially but the non-treatment group was able to catch up after surgery. The authors concluded that PSOT has no lasting positive effect on weight gain or growth in patients with oral clefts. A group of British researchers also attempted to substantiate this controversial claim by conducting their own randomized clinical trial. 50 infants with CUCLP and isolated cleft palate were included in the study [34]. Half of infant with CUCLP were randomly assigned to wear an active orthopedic device while half of infants with isolated cleft palate wore a passive device. The remainder did not receive any PSOT. These authors also found that PSOT was not associated with any improvement in feeding efficiency.

PSOT and Nasolabial Esthetics

The effect of PSOT nasolabial esthetics has historically been given less emphasis than its effects on maxillary growth. However, more studies have shed light on the subject during the past decade. In 2003, a group of Japanese researchers evaluated the effects of Hotz plate on nasolabial and alveolar changes [35]. They recruited 15 infants with CUCLP. All of them wore the appliance for 3 months. Palatal casts were used to assess alveolar changes while a 3-D optical scanner was used for nasolabial changes. They concluded that PSOT reduced the initial alveolar cleft width and made subsequent surgery easier, but had no effect on nasolabial morphology. The Dutchcleft studies also assessed the effects of PSOT on facial appearance in infants with oral clefts using facial photographs at the ages of 18 months, 4 years, and 6 years of age [36-37]. Professionals and laypersons rated these photographs according to the quality of nasolabial esthetic outcomes. At 18 months, the authors found no difference between the PSOT group and the non-treatment group. At 4 years of age, the PSOT group was found to be more attractive. However, this difference disappeared at 6 years of age. Only professionals thought that the PSOT group presented with a significantly better nasolabial outcome. The authors suggested that the difference found at 6 years of age might not be of any clinical significance since patients mainly interacted with laypersons in their ordinary lives.

Eurocleft studies evaluated the treatment satisfaction of patients at 17 years of age and their parents by questionnaires [38]. They found that patients were most dissatisfied with the appearance of the nose, lip, and facial profile. These levels of dissatisfaction with nasolabial esthetics were statistically higher than dissatisfaction expressed about speech. Parents also agreed with the patients. They were more dissatisfied with their child's nasal appearance, lip and the facial profile than with speech outcomes.

Nasoalveolar Molding

Although traditional PSOT reduces the initial severity of the alveolar cleft and created a better environment for surgical repair, it does not appear to have any long-term effect on facial growth or nasolabial esthetics [38]. It was the nasal esthetics that patients with oral cleft express their most dissatisfaction. Surgical correction of nasal asymmetry for patients with oral clefts has always been a challenge for any surgeons. Matsuo and colleagues reported the use of nasal conformers before the lip surgery to correct nasal asymmetry in patients with incomplete cleft lip and complete cleft lip with Simonart's band [39]. The principle of cartilage molding was first described by the same group of researchers in 1984 [40]. They demonstrated the use of a stent to mold the deformed auricular cartilage in infants to reduce the need for otoplasty. This was based on the findings that cartilages are moldable the first 6-8 weeks due to high level of estrogen transferred from mothers to infants at birth. This process may be necessary to relax ligaments, cartilages, and connective tissues enabling the fetus to pass through the birth canal. The level of estrogen decreases soon after birth, and consequently the ability to mold the cartilages diminishes. In 1993, Grayson, Cutting, and Wood briefly reported a PSOT technic that incorporated 2 nasal stents attached to an alveolar molding plate, to pre-surgically elongate a deficient columella and improve a nasal tip projection in a patient with a bilateral cleft lip and palate (BCLP) [41]. Several years later, Grayson and his colleagues described the same technique in greater detail [42]. They reported using this new technic in patients with BCLP as well as those with CUCLP. They utilized a nasal stent attached to an oral plate to pre-surgically mold both nose and alveolar bone, hence the term nasoalveolar molding (NAM). In their initial report, a nasal stent was made of acrylic resin, the same material used for making an oral plate. The stent was attached to an oral plate and extended inside the affected nostril. The principal treatment goals in patients with UCLP were not only to reduce the initial cleft size but also to improve nasal symmetry. In patients with BCLP, pre-surgical NAM aims at retracting the premaxilla palatally and elongating the columella.

The technique has gained in popularity over the past decade as it has allowed surgeons to consistently obtain better nasolabial outcomes. According to an international survey on treatment of cleft lip and palate published in 2008, 38% of the respondents performed NAM occasioanally[9]. A more recent survey published in 2012 reported that 37% of the U.S. cleft teams, provided NAM in their centers [43]. Interestingly, this survey also reported that infants who received NAM had larger cleft widths than those who did not.

NAM vs. Hotz Plate

Bennun and colleagues published a 6-year follow-up study comparing nasal features between a group of 44 infants with CUCLP who received pre-surgical NAM and a group of 47 infants with CUCLP who had McNeil-type PSOT [44]. The same surgeon performed the surgery in all patients. The authors reported that the NAM group had a better nostril symmetry and a significantly increased columellar length. They also suggested NAM as a way to reduce the need for surgical revision of the nose.Maull and colleagues conducted a similar study but with a smaller sample size. They compared a group of 10 randomly selected infants with CUCLP who underwent NAM pre-surgically with a group of 10 infants with CUCLP who received a conventional PSOT [45]. Nasal impressions were taken for symmetry assessment. They reported that NAM significantly increases the symmetry of the nose. The increase in symmetry was maintained into early childhood. Nakamura and colleagues compared 15 Japanese patients with CUCLP who were treated with NAM to 15 Japanese patients with CUCLP who used only Hotz plates [46]. They followed up the patients for 1-5 years. The authors concluded that the NAM group demonstrated a better the nostril height and width ratio than the Hotz plate group. Sasaki and colleagues conducted a similar study with a similar sample size in 28 Japanese patients with CUCLP [47]. 13 patients had NAM pre-surgically, and 15 received Hotz plate. The authors found that NAM improved the nasal form and decreased palatal cleft gap more effectively than Hotz plate. They also reported that the size of the palatal cleft correlated with the esthetic outcome of the nose, following surgery.

Pre-Surgical Effects of NAM

Nasal morphology was assessed in 10 infants with CUCLP treated with NAM [48]. Digital stereophotogrammetry was used to capture 3-D facial image before NAM and after NAM. The authors concluded that the use of NAM improved bilateral nasal symmetry before surgery. The same group of researchers also compared facial morphology in 15 patients (37 week old) with UCLP following NAM and cheiloplasty to 10 matched control subjects [49]. They reported that facial forms in patients who received NAM were similar to those from the non-cleft control subjects. Jaeger and colleagues studied 11 infants with CUCLP who were treated by NAM pre-surgically[50]. Before and after NAM conditions were assessed. The authors reported a significant reduction of the cleft size, in addition to a significant improvement of nasal symmetry and nostril shape. Ezzat and colleagues evaluated the effects of pre-surgical NAM before cheiloplasty in 20 infants with CUCLP by using intraoral and extraoral casts [51]. They also concluded that NAM decreased the initial alveolar cleft size while permitting an increase in posterior maxillary arch width. It also increased nasal symmetry by creating more symmetrical nostril heights and widths. Kecik and Enacaralso studied the effects of pre-surgical NAM [52]. 22 Turkish patients with CUCLP were included this prospective study. Maxillary and nasal models taken before and after NAM were measured. The authors reported a significant reduction in cleft size and positive changes in nasal features. Gomez and colleagues analyzed nasal changes before and after pre-surgical NAM in 30 patients with CUCLP [53]. Photogrammetric analysis was performed on nasal casts taken before and after NAM. The results showed a significant reduction of columellar deviation along with an increase in nasal tip projection that resulted in an improved nasal form prior to lip and nose surgery. The findings from these recent studies indicated that NAM is an effective treatment to pre-surgically improve nasal asymmetry.

Post-Surgical Nasal Changes

A team of Taiwanese researchers evaluated progressive changes of nasal symmetry in 57 Taiwanese infants with UCLP who underwent pre-surgical NAM [54]. They used basilar facial pictures taken before NAM, before cheiloplasty, 1 month after cheiloplasty, and 1 year of age to assess nostril forms. The authors concluded that there was a significantly improved nasal symmetry after NAM. The nose was further corrected to symmetry after primary cheiloplasty. However, there was some postsurgical relapse of nostril shape in width, height, and angle of columella at 1 year of age. Another Taiwanese research team conducted a similar study but extended the follow-up time to 3 years with a smaller sample size [55]. 25 consecutive Taiwanese infants with CUCLP participated in this study. They concluded that the relapsed nostril was only observed in the first year postoperatively but became stable afterward. They also explained that the relapse was due to a differential growth between nostrils of the cleft and non-cleft sides. Another research team from the same Taiwanese institution recommended a surgical overcorrection on the affected side of the nose and the postoperative use of nasal conformers for 12 months to prevent the relapse [56]. Barillas and his colleagues conducted a retrospective study assessing nasal symmetry of 9-year-old patients who had been treated for CUCLP during infancy [57]. 15 infants with CUCLP were treated with pre-surgical NAM and surgery whereas 10 patients only had surgery. The authors concluded that a greater degree of nasal symmetry could be observed in patients from the NAM group and was maintained into preadolescence.

NAM in Different Racial Groups

The favorable effects of pre-surgical NAM have been widely reported in Caucasian populations [41-42,45,48-50]. Similar reports have been also made in Asian and Turkish populations as well [46-47,52]. In addition to a previous study from Argentina [44], a team of researchers from Columbia evaluated the efficacy of pre-surgical NAM on nasal esthetics in a sample of Columbian patients with CUCLP [58]. 17 patients with CUCLP treated with NAM took part in the study. Facial impressions were taken before treatment, before primary rhinocheiloplasty, and before palatoplasty. The authors concluded that pre-surgical NAM effectively improved nasal tip projection, elevated the depressed alar cartilage, and reduced columellar deviation prior to surgery in Columbians.

Variation of Nam Technics

Since NAM was first reported, several authors have reported their modifications of the NAM technic. Grayson and his colleagues recommended that nasal molding should start when the alveolar cleft is reduced to less than 6 mm [42]. Suri and Thompson reported a technic described as "A modified muscle-activated maxillary orthopedic appliance for NAM" in infants with CUCLP [59]. A team of Japanese clinicians reported using a cobalt-chrome wire (1mm in diameter) to construct a nasal stent enabling a better and easier control of the force and direction of the stent[60]. Figuero and Polley used a nasal stent made of stainless steel. They started nasal and alveolar molding at the same time [61]. Liao and her colleagues investigated the optimal timing to start nasal molding [62]. 63 Taiwanese patients with CUCLP participated in the study. 31 received the Grayson's NAM, which starts nasal molding when an alveolar cleft width is reduced to 5 mm. The remainder underwent treatment with the Figueroa's NAM, which starts nasal and alveolar molding at the same time. Pre-treatment and post-treatment facial photographs were used to compare the difference. The findings showed that both technics produced clinically similar outcomes.

Digital imaging technology has played an increasingly important role in both medicine and dentistry. It is no surprise that clinicians began to incorporate the technology into daily clinical practice. A research team from China reported using a 3-D laser scanning to obtain a digital maxillary model for the fabrication of NAM appliances [63]. The appliances were constructed with a rapid prototyping technique, which allowed the complete set of appliances to be made at one time.

The NAM technic that I currently use consists of an oral plate made of acrylic resin typically used for making orthodontic retainers, and a nasal stent made of a stainless steel wire of 1 mm in diameter. The nasal stent is adhered to the labial flange of the oral plate. The tip of the nasal stent is covered with soft denture liner to prevent tissue irritation (Figures 1-2). The NAM plate is retained with denture adhesive. Parents are also asked to perform lip taping with a thin strip of skin colored tape to approximate the lip segments. This taping technic is designed to minimize visibility of the oral clefts and allow the patients to appear more normal even before surgery (Figure 3). Patients are usually seen once a week so that the appliance can be adjusted to make treatment progress. The alveolar and nasal molding starts simultaneously. The treatment usually lasts between 3-5 months depending on the severity of the condition. A series of patients was presented to demonstrate the outcomes achieved with the NAM technic (Figures 4-6). In patients with incomplete unilateral cleft lip with or without cleft palate, nasal springs were used to correct the nasal asymmetry. Further discussion on this topic will be presented later in the chapter.

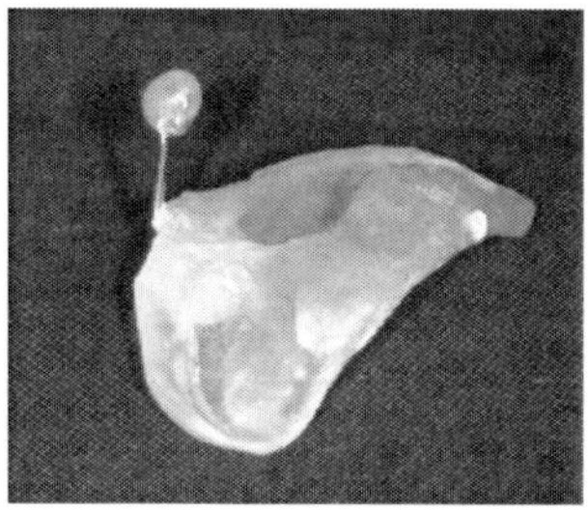

Figure 1. NAM appliance for infants with unilateral cleft lip and palate.

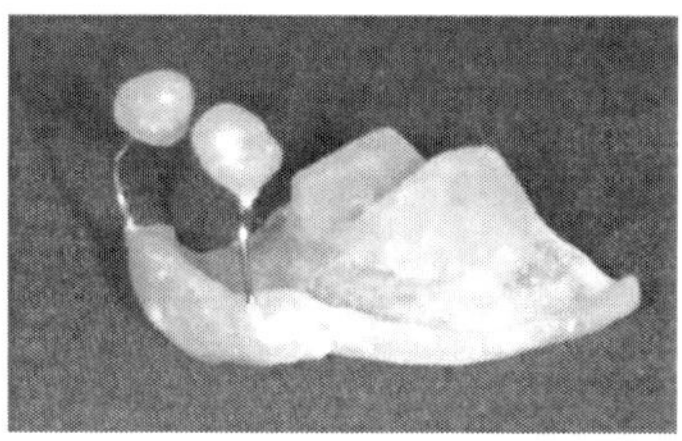

Figure 2. NAM appliance for infants with bilateral cleft lip and palate.

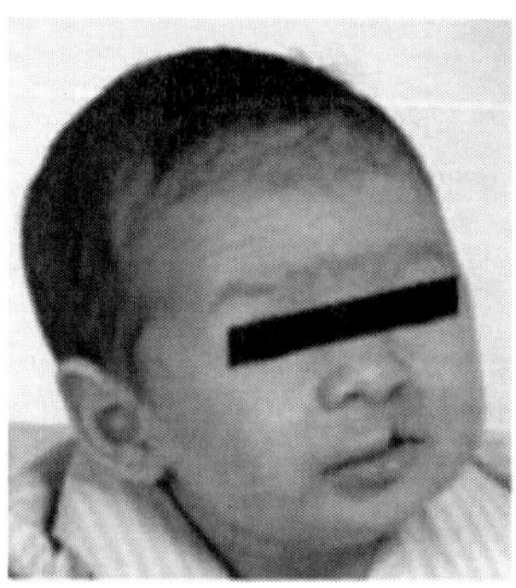

Figure 3. An infant with a complete unilateral cleft lip and palate with NAM appliance and lip taping. This taping technic reduces the visibility of the oral clefts.

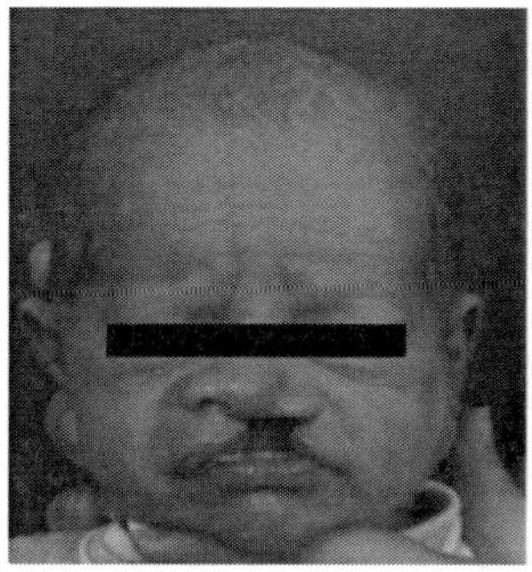

Figure 4A. An infant with CUCLP before NAM.

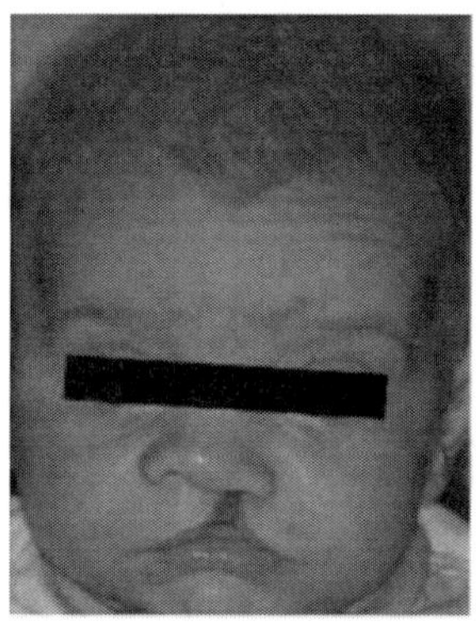

Figure 4B. The patient exhibited an improved nasal symmetry after NAM. The lip segments were approximated.

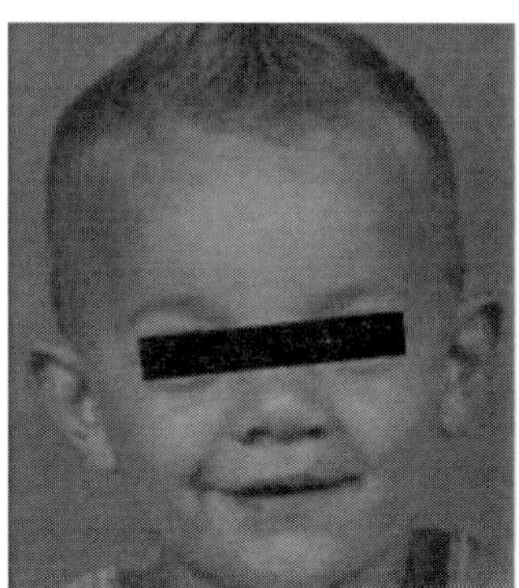

Figure 4C. The surgery further corrected the nasal form. The patient exhibited a good nasal symmetry and lip form.

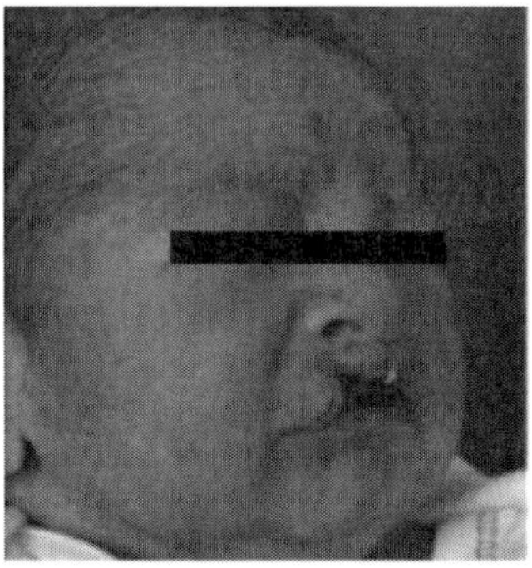

Figure 5A. A patient with BCLP.

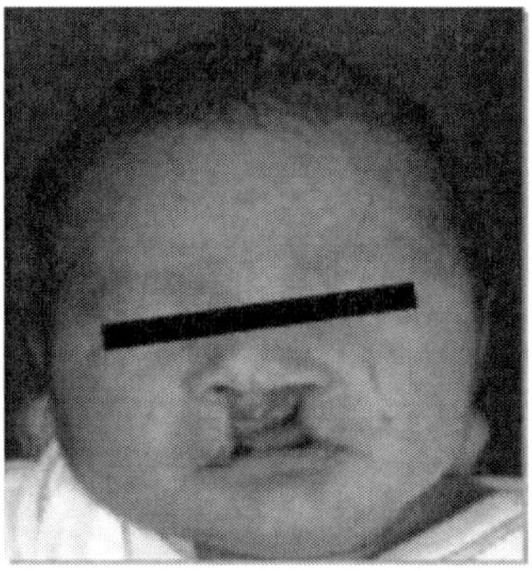

Figure 5B. A frontal view.

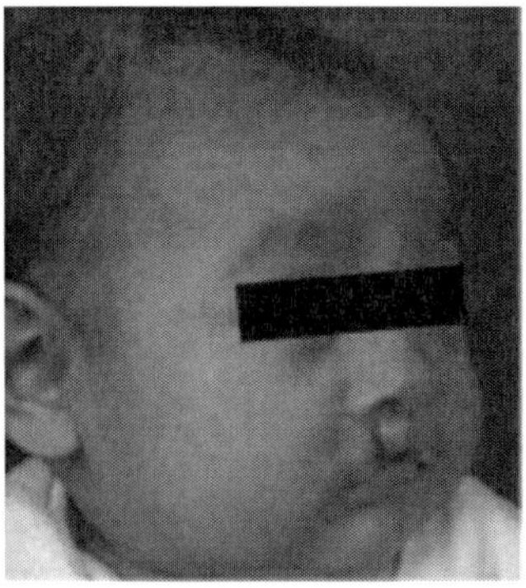

Figure 5C. The deficient columella was elongated presurgically by NAM. The nasal tip projection was significantly increased.

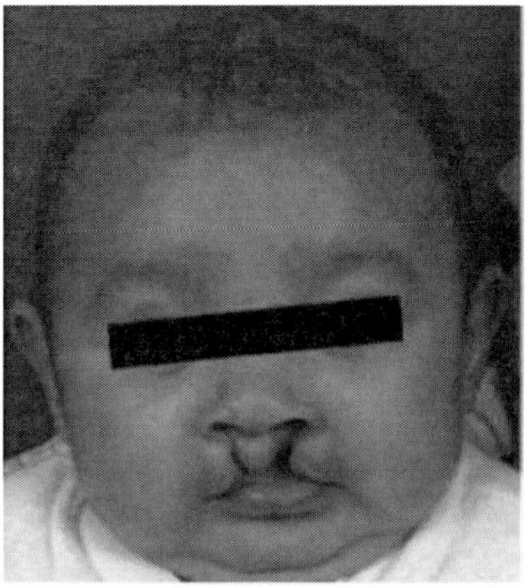

Figure 5D. The lip segments were approximated. The nasal symmetry was improved significantly.

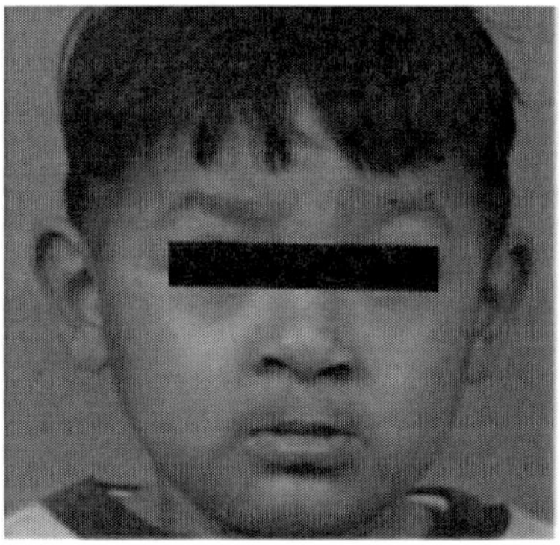

Figure 5E. After surgery, the nose has a normal form and projection.

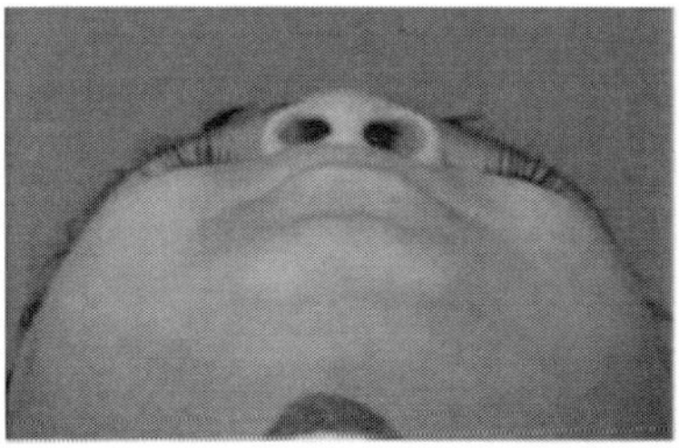

Figure 5F. The patient obtained a normal columellar length after surgery.

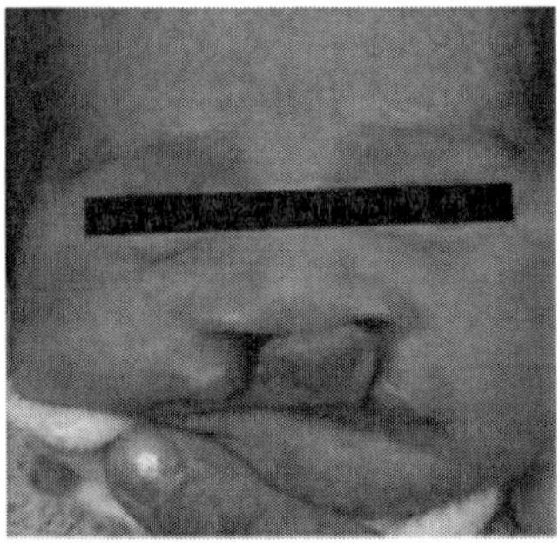

Figure 6A. A patient with complete bilateral cleft lip and palate.

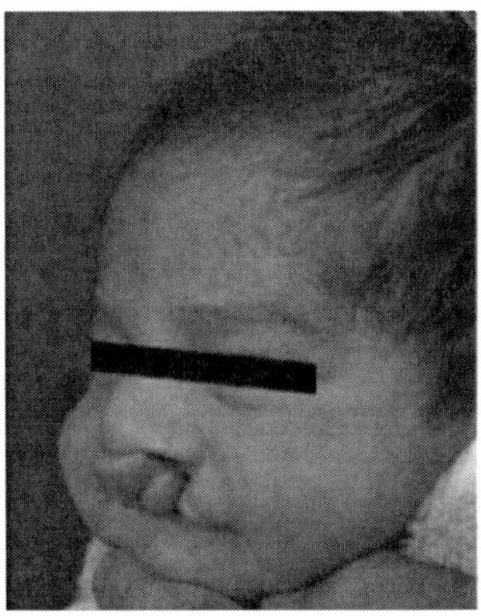

Figure 6B. Profile view.

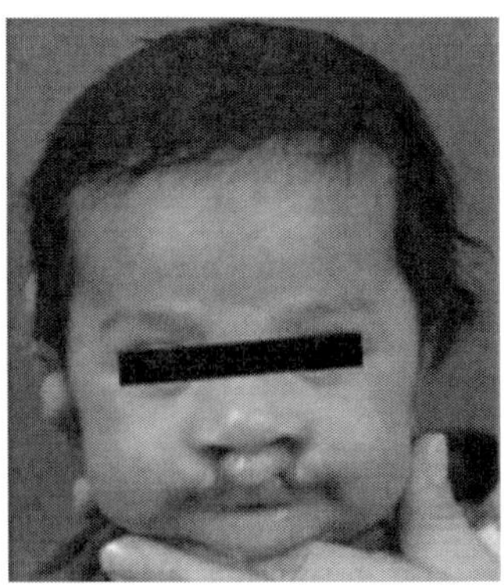

Figure 6C. The premaxilla was retracted after NAM. Nasal width was reduced. All lip segment were approximated.

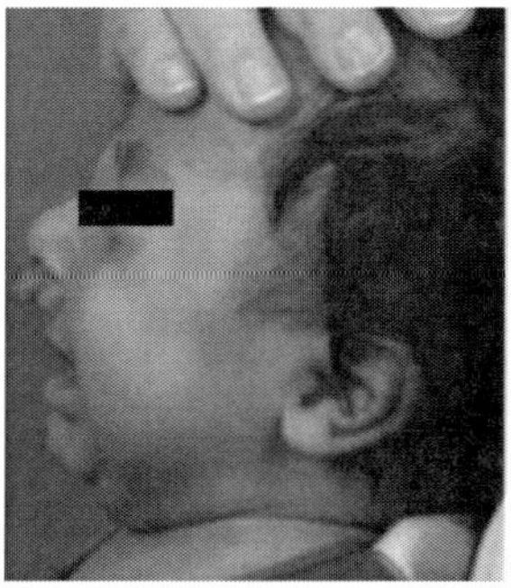

Figure 6D. The columella was elongated after NAM. The nasal tip projection was significantly improved.

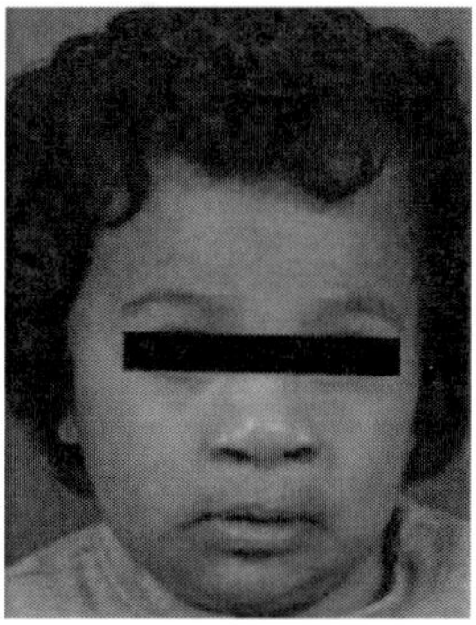

Figure 6E. After surgery, the patient has an esthetically pleasing nasolabial structures.

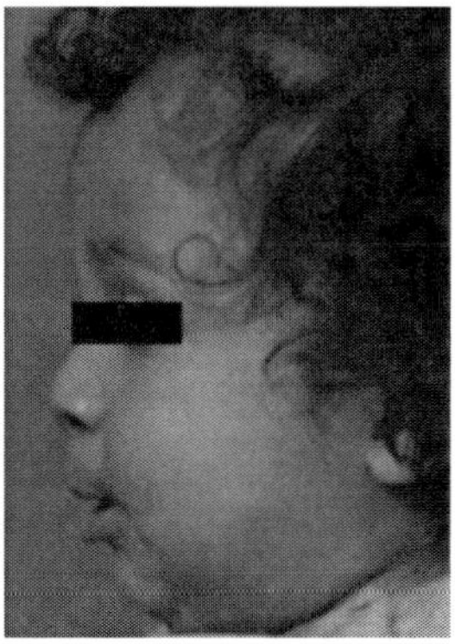

Figure 6F. The patient demonstrated a good nasal tip projection and columella from a profile view.

NAM in Patients with Bilateral Cleft Lip and Palate (BCLP)

Spengler and colleagues evaluated the outcomes of pre-surgical NAM in 8 patients with BCLP using intraoral and extraoral casts [64]. They reported that NAM pre-surgically improved the nasal asymmetry, increased nasal tip projection, and retracted the protruded premaxillary segment. Lee and colleagues conducted a retrospective study of 26 consecutive patients with BCLP [65]. 13 had a surgery only whereas the remainder had NAM followed by lip surgery and primary retrograde nasal correction. Columellar length was measured before treatment and at 3 years of age. The number of nasal operations was recorded up to 9 years of age. The authors found that the columellar length was significantly greater in the NAM group, similar to non-cleft children. All patients in the surgery only group needed some nasal revision while no patient in the NAM required any secondary nasal surgery.

A team of researcher from Taiwan assessed the progressive changes of columella length and nasal growth after NAM and primary cheiloplasty[66]. 32 consecutive infants with BCLP were included in this study. All of them were treated with NAM. Standard 1:1 basilar photographs were taken before and after NAM, 1 week after cheiloplasty, and annually for 3 years. The findings indicated that columellar length was significantly increased after NAM and was further improved after primary cheiloplasty. The authors further reported a relative relapse in columellar length due the differential growth between the columella and the rest of the nose in the first and second year post-surgically.

Meazzini and colleagues evaluated nasal esthetics of children with BCLP, treated with pre-surgical NAM and surgery at 5 years of age [67]. They compared 2 groups of children with BCLP to non-cleft children.18 patients were treated with NAM and surgery. 18 patients underwent surgery only. The authors reported that the columella length, nasal tip angle, and protrusion were greatly improved and were close to normal. However, the nasolabial angle and interalar distances were larger than normal in both groups. Suri and colleagues compared infants with BCLP who received pre-surgical NAM with those who had traditional PSOT only [68]. Cheiloplasty was performed by a single surgeon. 29 patients received NAM whereas 17 infants underwent only PSOT. Nasoalveolar changes were assessed prior to lip surgery. The authors reported that infants who underwent NAM had longer columella and better-aligned alveolar segments than those who received only PSOT.

The findings from these recent studies indicated that pre-surgical NAM increased columella length and nasal tip projections. It also reduced premaxillary protrusion, easing the process of lip surgery. The columellar length was maintained post-surgically. As a result, secondary nasal surgery was usually not needed for patients who received pre-surgical NAM.

NAM and Gingivoperiosteoplasty (GPP)

A treatment protocol consisting of pre-surgical NAM, combined lip and primary nasal pair and GPP, is not only well known but also controversial [42]. Santiago and colleagues reported that 60% of patients with CUCLP who underwent GPP at the time of lip repair do not require a subsequent alveolar bone graft [69]. Dec and colleagues reported a success rate of alveolar bone formation following primary bilateral GPP to be around 52% [70]. However, little is known regarding the effect of this treatment protocol on the maxillary growth. Lee and colleagues conducted retrospective study of 20 consecutive patients with a CUCLP [71]. 10 patients received NAM and had GPP performed at lip closure; 10 patients had NAM but no GPP. A single surgeon performed all surgical procedures. Cephalometric radiographs were evaluated before bone grafting and at pre-puberty age. The results suggested that mid-facial growth in sagittal or vertical planes (up to the age of 9 to 13 years) were not affected by NAM and GPP. But when a similar study was conducted with a much larger sample size, Hsieh and colleagues found that GPP had significantly negative effects on the maxillary growth [72]. They studied 62 consecutive patients with CUCLP. All the patients received NAM treatment before cheiloplasty. 26 patients had GPP performed at the time of cheiloplasty whereas 36 patients did not. Cephalometry was used to evaluate the facial growth at 5 years of age in both groups of patients. The authors found that sagittal growth of the maxilla was affected more adversely in the GPP group than in the non-GPP group at the age of 5 years.

Nasal Spring

Although NAM has been shown to consistently improve nasolabial esthetics. The technique is often criticized for requiring an experienced clinician to perform, and a significant compliance from parents. I designed a nasal spring to mold alar cartilage to correct congenital nasal asymmetry on infants born with incomplete unilateral cleft lip with or without cleft palate. The spring is simpler in design and construction. It does not require an oral plate for adherence that a traditionally designed NAM plate does. Therefore, this technique is less demanding for parents and not as labor intensive for clinicians. A nasal spring is made of stainless steel wire bent into a horseshoe shape (Figure 7). The spring has 2 arms. The arm that goes inside the affected nostril is bent anteriorly so that it applies pressure on the alar cartilage. The other arm is bent distally so that will not block the air passage. The spring is secured over the lip by taping (Figure 8). The improvement of nasal symmetry can be observed after wearing the nasal spring for 3 months (Figures 9A-9C). As most recent studies showed a lack oforthodontic benefits from a traditional PSOT, perhaps a nasal molding device may be all we need to achieve a better nasal outcome. An improvement or modification of the nasal spring is still needed so that it can be used effectively in patients with CUCLP.

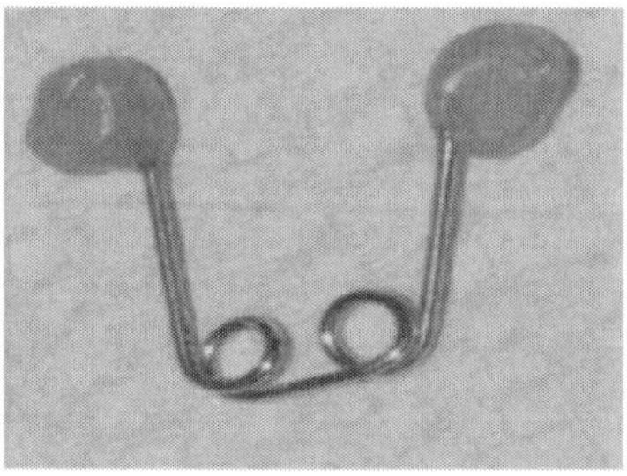

Figure 7. Nasal spring.

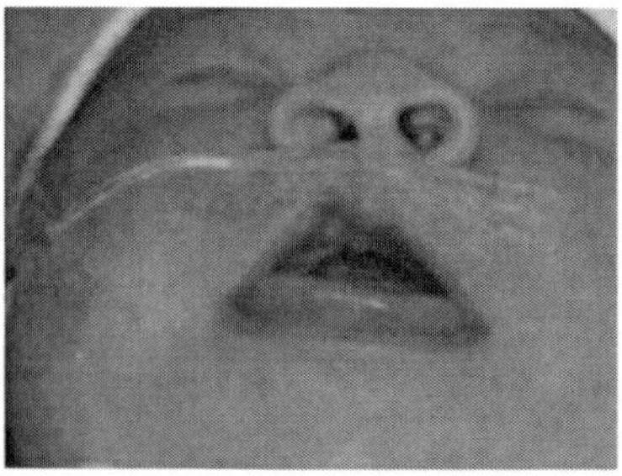

Figure 8. A nasal spring was placed on the lip and secured by tape.

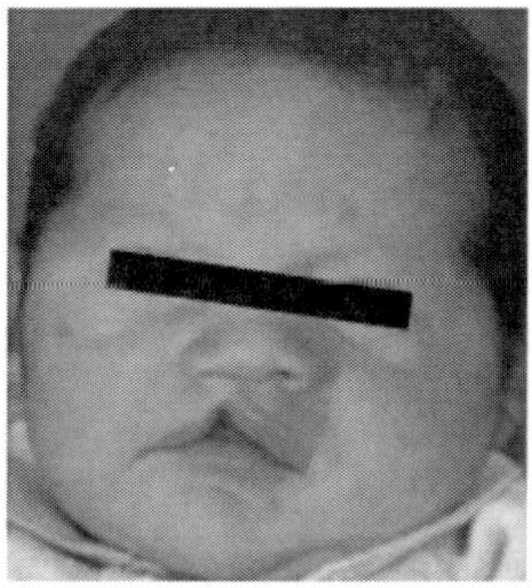

Figure 9A. An infant with incomplete cleft lip with moderate nasal deformity secondary to oral cleft.

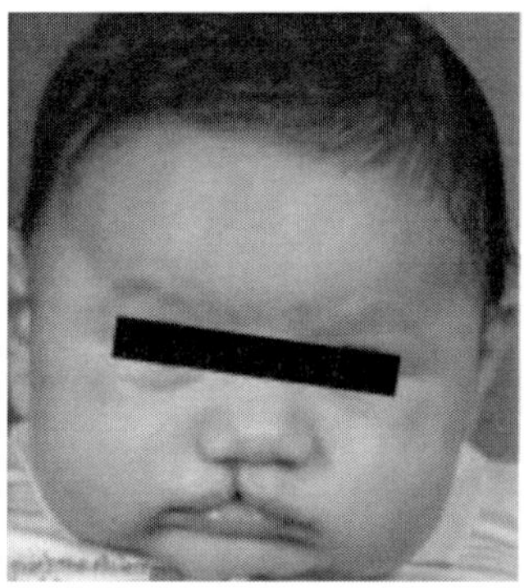

Figure 9B. Nasal symmetry was achieved after the patient used a nasal spring for 3 months.

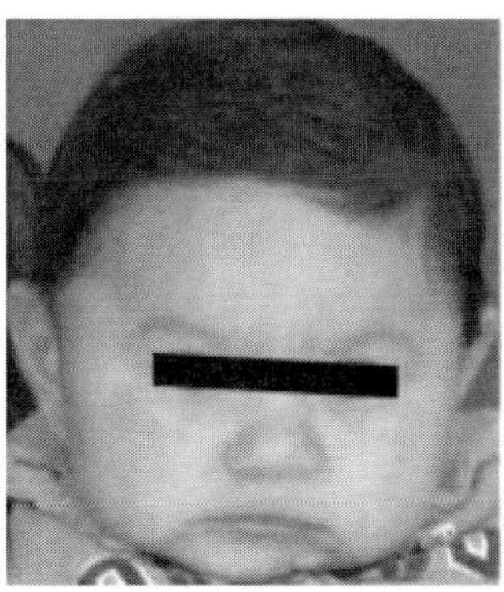

Figure 9C. Patient after surgery.

Conclusion

The treatment goals for NAM have gone beyond the traditional goals of PSOT to put more emphasis on nasolabial esthetics. Pre-surgical NAM put more focus on achieving a superior long-term nasolabial esthetics, and minimizing or avoiding any future nasolabial surgical revision. In light of the findings presented in this chapter, it can be concluded that NAM has brought about a paradigm shift in PSOT for infants with oral cleft.

References

[1] McNeil, CK. Orthodontic procedures in the treatment of congenital cleft palate. *Dent Rec*. 1950, 70, 126–132.

[2] Hotz, M; Gnoinski, W. Comprehensive care of the cleft lip and palate children at Zurich University: a preliminary report. *Am J Orthod.*1976, 70, 481–504.

[3] Latham, RA. Orthodontic advancement of the cleft maxillary segment: a preliminary report. *Cleft Palate J*. 1980, 17, 227–233.

[4] Millard, DR; Latham, RA. Improved primary surgical and dental treatment of clefts. *PlastReconstr Surg*. 1990, 86, 856–871.

[5] Millard, DR; Latham, R; Huifen, X; Spiro, S; Morovic, CG. Cleft lip and palatetreated by presurgical orthopedics, gingivoperiosteoplasty, and lip adhesion (POPLA) compared with previous lip adhesion method: a preliminary study of serial dental casts. *PlastReconstr Surg*. 1999, 103, 1630–1644.

[6] Berkowitz, S; Mejia, M; Bystrik, A. A comparison of the effects of the Latham-Millard procedure with those of a conservative treatment approach for dental occlusion and facial aesthetics in unilateral and bilateral complete cleft lip and palate: part I. Dental occlusion. *PlastReconstr Surg*. 2004, 113(1), 1-18.

[7] Matic, DB; Power, SM. The effects of gingivoperiosteoplasty following alveolar molding with a pin-retained Latham appliance versus secondary bone grafting on midfacial growth in patients with unilateral clefts. *PlastReconstr Surg*. 2008, 122(3), 863-70.

[8] Power, SM; Matic, DB. Gingivoperiosteoplasty following alveolar molding with a Latham appliance versus secondary bone grafting: the

effects on bone production and midfacial growth in patients with bilateral clefts. *PlastReconstr Surg*. 2009, 124(2), 573-82.

[9] Weinfeld, AB; Hollier, LH; Spira, M; Stal, S. International trends in the treatment of cleft lip and palate. *Clin Plast Surg*. 2005, 32(1), 19-23

[10] Ross, RB. Treatment variables affecting facial growth in complete unilateral cleft lip and palate. Part II. Presurgicalorthopaedics. *Cleft Palate Craniofac J*. 1987, 24, 24-32.

[11] Shaw, WC; Asher-McDade, C; Brattstrom, V; Dahl, E; Mars, M; McWilliam, J; Mølsted, K; Plint, DA; Prahl-Andersen, B; Semb, G. The RPS. A six-center international study of treatment outcome in patients with clefts of the lip and palate: Part 1. Principles and study design. *Cleft Palate Craniofac J*. 1992, 29, 393–397.

[12] Mølsted, K; Asher-McDade, C; Brattstrom, V; Dahl, E; Mars, M; McWilliam, W; Plint, DA; Prahl-Andersen, B; Semb, G; Shaw, WC. The RPS: a six-center international study of treatment outcome in patients with clefts of the lip and palate: Part 2. Craniofacial form and soft tissue profile. *Cleft Pal Craniofac J*. 1992, 29, 398–404.

[13] Mars, M; Asher-McDade, C; Brattstrom, V; Dahl, E; McWilliam, J; Mølsted, K; Plint, DA; Prahl-Andersen, B; Semb, G; Shaw, WC; et al. A six-center international study of treatment outcome in patients with clefts of the lip and palate: Part 3. Dental arch relationships. *Cleft Palate Craniofac J*. 1992, 29, 405–408.

[14] Mølsted, K; Dahl, E; Brattström, V; McWilliam, J; Semb, G. A six-center international study of treatment outcome in patients with clefts of the lip and palate: evaluation of maxillary asymmetry. *Cleft Palate Craniofac J*. 1993, 30(1), 22-8.

[15] Semb, G; Brattström, V; Mølsted, K; Prahl-Andersen, B; Shaw, WC. The Eurocleft study: intercenter study of treatment outcome in patients with complete cleft lip and palate. Part 1: introduction and treatment experience. *Cleft Palate Craniofac J*. 2005, 42(1), 64-8.

[16] Brattström, V; Mølsted, K; Prahl-Andersen, B; Semb, G; Shaw, WC. The Eurocleft study: intercenter study of treatment outcome in patients with complete cleft lip and palate. Part 2: craniofacial form and nasolabial appearance. *Cleft Palate Craniofac J*. 2005, 42(1), 69-77.

[17] Mølsted, K; Brattström, V; Prahl-Andersen, B; Shaw, WC; Semb, G. The Eurocleft study: intercenter study of treatment outcome in patients with complete cleft lip and palate. Part 3: dental arch relationships. *Cleft Palate Craniofac J*. 2005, 42(1), 78-82

[18] Long, RE Jr; Hathaway, R; Daskalogiannakis, J; Mercado, A; Russell, K; Cohen, M; Semb, G; Shaw, W. The Americleft study: an inter-center study of treatment outcomes for patients with unilateral cleft lip and palate part 1. Principles and study design. *Cleft Palate Craniofac J.* 201, 48(3), 239-43.

[19] Hathaway, R; Daskalogiannakis, J; Mercado, A; Russell, K; Long, RE Jr; Cohen, M; Semb, G; Shaw, W. The Americleft study: an inter-center study of treatment outcomes for patients with unilateral cleft lip and palate part 2. Dental arch relationships. *Cleft Palate Craniofac J.* 2011, 48(3), 244-51.

[20] Daskalogiannakis, J; Mercado, A; Russell, K; Hathaway, R; Dugas, G; Long, RE Jr; Cohen, M; Semb, G; Shaw, W. The Americleft study: an inter-center study of treatment outcomes for patients with unilateral cleft lip and palate part 3. Analysis of craniofacial form. *Cleft Palate Craniofac J.* 2011, 48(3), 252-8.

[21] Adali, N; Mars, M; Petrie, A; Noar, J; Sommerlad, B. Presurgical orthopedics has no effect on archform in unilateral cleft lip and palate. *Cleft Palate Craniofac J.* 2012, 49(1), 5-13.

[22] Mishima, K; Sugahara, T; Mori, Y; Sakuda, M. Three-dimensional comparison between the palatal forms in complete unilateral cleft lip and palate with and without Hotz plate from cheiloplasty to palatoplasty. *Cleft Palate Craniofac J.* 1996, 33, 312–317

[23] Mishima, K; Mori, Y; Sugahara, T; Minami, K; Sakuda, M. Comparison between palatal configurations in UCLP infants with and without a Hotz plate until four years of age. *Cleft Palate Craniofac J.* 2000, 37(2), 185-90.

[24] Prahl, C; Kuijpers-Jagtman, AM; van't Hof, MA; Prahl-Andersen, B. A randomised prospective clinical trial into the effect of infant orthopaedics on maxillary arch dimensions in unilateral cleft lip and palate (Dutchcleft). *Eur J Oral Sci.* 2001, 109, 297–305.

[25] Prahl, C; Kuijpers-Jagtman, AM; van't Hof, MA; Prahl-Andersen, B. A randomized prospective clinical trial into the effect of infant orthopedics in UCLP: prevention of collapse of the alveolar segments (Dutchcleft). *Cleft Palate Craniofac J.* 2003, 40, 337–342.

[26] Bongaarts, CAM; Kuijpers-Jagtman, AM; van't Hof, MA; Prahl-Andersen, B. The effect of infant orthopedics on the occlusion of the deciduous dentition in children with complete unilateral cleft lip and palate (Dutchcleft). *Cleft Palate Craniofac J.* 2004, 41, 633–641.

[27] Bongaarts, CAM; van't Hof, MA; Prahl-Andersen, B; Dirks, IV; Kuijpers-Jagtman, AM. Infant orthopedics (IO) has no effect on maxillary arch dimensions in the deciduous dentition of children with complete unilateral cleft lip and palate (Dutchcleft). *Cleft Palate Craniofac J.* 2006, 43, 665–672.

[28] Bongaarts, CA; Prahl-Andersen, B; Bronkhorst, EM; Prahl, C; Ongkosuwito, EM; Borstlap, WA; Kuijpers-Jagtman, AM. Infant orthopedics and facial growth in complete unilateral cleft lip and palate until six years of age (Dutchcleft). *Cleft Palate Craniofac J.* 2009, 46(6), 654-63.

[29] Hardin-Jones, MA; Chapman, KL; Wright, J; Halter, KA; Schulte, J; Dean, JA; Havlik, RJ; Goldstein, J. The impact of early palatal obturation on consonant development in babies with unrepaired cleft palate. *Cleft Palate Craniofac J.* 2002, 39(2), 157-63.

[30] Konst, EM; Rietveld, T; Peters, H; Weersink-Braks, H. Use of perceptual evaluation instrument to assess the effects of infant orthopedics on the speech of toddlers with cleft lip and palate. *Cleft Palate Craniofac J.* 2003a, 40, 597–605.

[31] Konst, EM; Rietveld, T; Peters, H; Prahl-Andersen, B. Phonological development of toddlers with unilateral cleft lip and palate who were treated with and without infant orthopedics: a randomized clinical trial. *Cleft Palate Craniofac J.* 2003b, 40, 32–39.

[32] Konst, EM; Rietveld, T; Peters, H; Kuijpers-Jagtman, AM. Language skills of young children with unilateral cleft lip and palate following infant orthopedics: a randomized clinical trial. *Cleft Palate Craniofac J.* 2003c, 40, 356–362.

[33] Prahl, C; Kuijpers-Jagtman, AM; van't Hof, MA; Prahl-Andersen, B. Infant orthopedics in UCLP: effect on feeding, weight and length: a randomized clinical trial (Dutchcleft). *Cleft Palate Craniofac J.* 2005, 42, 171–177.

[34] Masarei, AG; Wade, A; Mars, M; Sommerlad, BC; Sell, D. A randomized control trial investigating the effect of presurgical orthopedics on feeding in infants with cleft lip and/or palate. *Cleft Palate Craniofac J.* 2007, 44(2), 182-93.

[35] Yamada, T; Mori, Y; Mishima, K; Sugahara, T. Nasolabial and alveolar morphology following presurgicalorthopaedic treatment in complete unilateral clefts of lip, alveolus and palate. *J Craniomaxillofac Surg.* 2003, 31(6), 343-7.

[36] Bongaarts, CAM; Prahl-Andersen, B; Bronkhorst, EM; Spauwen, PHM; Mulder, JW; Vaandrager, JM; Kuijpers-Jagtman, AM. Effect of infant orthopedics on facial appearance of toddlers with complete unilateral cleft lip and palate (DUTCHCLEFT). *Cleft Palate Craniofac J.* 2008b, 45, 407–413.

[37] Prahl, C; Kuijpers-Jagtman, AM; van't Hof, MA; Prahl-Andersen, B. Infant orthopedics and facial appearance: a randomized clinical trial (Dutchcleft). *Cleft Pal Craniofac J.* 2006, 43, 659–664.

[38] Semb, G; Brattstrom, V; Mølsted, K; Prahl-Andersen, B; Zuurbier, P; Rumsey, N; Shaw, WC. The Eurocleft study: longitudinal follow-up of patients with complete cleft lip and palate: Part 4. Relationship between treatment outcome, patient/parent satisfaction and the burden of care. *Cleft Palate Craniofac J.* 2004, 42, 83–92.

[39] Matsuo, K; Hirose, T; Otagiri, T; Norose, N. Repair of cleft lip with nonsurgical correction of nasal deformity in the early neonatal period. *PlastReconstr Surg.* 1989, 83(1), 25-31.

[40] Matsuo, K; Hirose, T; Tonomo, T. Nonsurgical correction of congenital auricular deformities in the early neonate: a preliminary report. *PlastReconstr Surg.* 1984, 73, 38–50.

[41] Grayson, BH; Cutting, CB; Wood, R. Preoperative columella lengthening in bilateral cleft lip and palate. *PlastReconstr Surg.* 1993, 92, 1422–1423.

[42] Grayson, BH; Santiago, P; Brecht, L; Cutting, CB. Presurgicalnasoalveolar molding in patients with cleft lip and palate. *Cleft Palate Craniofac J.* 1999, 36, 486–498.

[43] Sischo, L; Chan, JW; Stein, M; Smith, C; van Aalst, J; Broder, HL. Nasoalveolar Molding: Prevalence of Cleft Centers Offering NAM and Who Seeks It. *Cleft Palate Craniofac J.* 2012, 49(3), 270-5.

[44] Bennun, RD; Perandones, C; Sepliarsky, VA; Chantiri, SN; Aguirre, MI; Dogliotti, PL. Nonsurgical correction of nasal deformity in unilateral complete cleft lip: a 6-year follow-up. *PlastReconstr Surg.* 1999, 104(3), 616-30.

[45] Maull, DJ; Grayson, BH; Cutting, CB; Brecht, LL; Bookstein, FL; Khorrambadi, D; Webb, JA; Hurwitz, DJ. Long-term effects of nasoalveolar molding on three-dimensional nasal shape in unilateral clefts. *Cleft Palate Craniofac J.* 1999, 36, 391–397.

[46] Nakamura, N; Sasaguri, M; Nozoe, E; Nishihara, K; Hasegawa, H; Nakamura, S. Postoperative nasal forms after presurgicalnasoalveolar molding followed by medial-upward advancement of nasolabial

components with vestibular expansion for children with unilateral complete cleft lip and palate. *J Oral Maxillofac Surg*. 2009, 67(10), 2222-31.

[47] Sasaki, H; Togashi, S; Karube, R; Yanagawa, T; Nakane, S; Tabuchi, K; Ishibashi, N; Shinya, Y; Ito, H; Yamagata, K; Onizawa, K; Adachi, K; Sekido, M; Bukawa, H. Presurgicalnasoalveolar molding orthopedic treatment improves the outcome of primary cheiloplasty of unilateral complete cleft lip and palate, as assessed by naris morphology and cleft gap. *J Craniofac Surg*. 2012, 23(6), 1596-601.

[48] Singh, GD; Levy-Bercowski, D; Santiago, PE. Three-dimensional nasal changes following nasoalveolar molding in patients with unilateral cleft lip and palate: geometric morphometrics. *Cleft Palate Craniofac J*. 2005, 42(4), 403-9.

[49] Singh, GD; Levy-Bercowski, D; Yáñez, MA; Santiago, PE. Three-dimensional facial morphology following surgical repair of unilateral cleft lip and palate in patients after nasoalveolar molding. *OrthodCraniofac Res*. 2007, 10(3), 161-6.

[50] Jaeger, M; Braga-Silva, J; Gehlen, D; Sato, Y; Zuker, R; Fisher, D. Correction of the alveolar gap and nostril deformity by presurgical passive orthodontia in the unilateral cleft lip. *Ann Plast Surg*. 2007, 59(5), 489-94.

[51] Ezzat, CF; Chavarria, C; Teichgraeber, JF, et al. Presurgicalnasoalveolar molding therapy for the treatment of unilateral cleft lip and palate: A preliminary study. *Cleft Palate Craniofac J*. 2007, 44, 8–12.

[52] Kecik, D; Enacar, A. Effects of nasoalveolar molding therapy on nasal and alveolar morphology in unilateral cleft lip and palate. *J Craniofac Surg*. 2009, 20, 2075–2080.

[53] Gomez, DF; Donohue, S; Figueroa, AA; Polley, JW. Nasal changes after presurgicalnasoalveolar molding (PNAM) in the unilateral cleft lip nose. *Cleft Palate Craniofac J*. 2012, 49(6), 689-700

[54] Pai, BC; Ko, EW; Huang, CS; Liou, EJ. Symmetry of the nose after presurgicalnasoalveolar molding in infants with uni- lateral cleft lip and palate: A preliminary study. *Cleft Palate Craniofac J*. 2005, 42, 658–663.

[55] Liou, EJ; Subramanian, M; Chen, PK; Huang, CS. The pro- gressive changes of nasal symmetry and growth after naso-alveolar molding: A three-year follow-up study. *PlastReconstr Surg*. 2004, 114, 858–864.

[56] Chang, CS; Por, YC; Liou, EJ; Chang, CJ; Chen, PK; Noordhoff, MS. Long-term comparison of four techniques for obtaining nasal symmetry

in unilateral complete cleft lip patients: a single surgeon's experience. *PlastReconstr Surg*. 2010, 126(4), 1276-84.

[57] Barillas, I; Dec, W; Warren, SM; Cutting, CB; Grayson, BH. Nasoalveolar molding improves long-term nasal symmetry in complete unilateral cleft lip-cleft palate patients. *PlastReconstr Surg*. 2009, 123, 1002–1006.

[58] López-Palacio, AM; Cerón-Zapata, AM; Gómez, DF; Dávila-Calle, AP; Ojalvo-Arias, MA. Nasal changes with nasoalveolar molding in Colombian patients with unilateral cleft lip and palate. *Pediatr Dent*. 2012, 34(3), 239-44.

[59] Suri, S; Tompson, BD. A modified muscle-activated maxillary orthopedic appliance for presurgicalnasoalveolar molding in infants with unilateral cleft lip and palate. *Cleft Palate Craniofac J*. 2004, 41(3), 225-9.

[60] Mitsuyoshi, I; Masahiko, W; Masayuki, F. Simple modified preoperative nasoalveolarmoulding in infants with unilateral cleft lip and palate. *Br J Oral Maxillofac Surg*. 2004, 42(6), 578-80.

[61] Figueroa, AA; Polley, JW. Orthodontics in cleft lip and palate management. In: Mathes S, Ventz V, eds. *Plastic Surgery*. Vol. 4, Pediatric Plastic Surgery.2nd ed. Philadelphia: Elsevier; 2006, 271–310.

[62] Liao, YF; Hsieh, YJ; Chen, IJ; Ko, WC; Chen, PK. Comparative outcomes of two nasoalveolar molding techniques for unilateral cleft nose deformity. *PlastReconstr Surg*. 2012, 130(6), 1289-95.

[63] Yu, Q; Gong, X; Wang, GM; Yu, ZY; Qian, YF; Shen, G. A novel technique for presurgicalnasoalveolar molding using computer-aided reverse engineering and rapid prototyping. *J Craniofac Surg*. 2011, 22(1), 142-6.

[64] Spengler, AL; Chavarria, C; Teichgraeber, JF; Gateno, J; Xia, JJ. Presurgicalnasoalveolar molding therapy for the treatment of bilateral cleft lip and palate: A preliminary study. *Cleft Palate Craniofac J*. 2006, 43(3), 321-8.

[65] Lee, CT; Garfinkle, JS; Warren, SM; Brecht, LE; Cutting, CB; Grayson, BH. Nasoalveolar molding improves appearance of children with bilateral cleft lip-cleft palate. *PlastReconstr Surg*. 2008, 122(4), 1131-7.

[66] Liou, EJ; Subramanian, M; Chen, PK. Progressive changes of columella length and nasal growth after nasoalveolar molding in bilateral cleft patients: a 3-year follow-up study.*PlastReconstr Surg*. 2007, 119(2), 642-8.

[67] Meazzini, MC; Rossetti, G; Morabito, A; Garattini, G; Brusati, R. Photometric evaluation of bilateral cleft lip and palate patients after primary columella lengthening. *Cleft Palate Craniofac J.* 2010, 47(1), 58-65.

[68] Suri, S; Disthaporn, S; Atenafu, EG; Fisher, DM. Presurgical presentation of columellar features, nostril anatomy, and alveolar alignment in bilateral cleft lip and palate after infant orthopedics with and without nasoalveolar molding.*Cleft Palate Craniofac J.* 2012, 49(3), 314-24.

[69] Santiago, PE; Grayson, BH; Gianoutsos, MP; Kwon, SM; Brecht, LE; Cutting, CB. Reduced need for alveolar bone grafting by pre-surgical orthopedics and primary gingivoperiosteoplasty. *Cleft Palate Craniofac J.* 1998, 35, 77–80.

[70] Dec, W; Shetye, PR; Davidson, EH; Grayson, BH; Brecht, LE; Cutting, CB; Warren, SM. Presurgicalnasoalveolar molding and primary gingivoperiosteoplasty reduce the need for bone grafting in patients with bilateral clefts. *J Craniofac Surg.* 2013, 24(1), 186-90.

[71] Lee, CT; Grayson, BH; Cutting, CB; Brecht, LE; Lin, WY. Prepubertalmidface growth in unilateral cleft lip and palate following alveolar molding and gingivoperiosteoplasty.*Cleft Palate Craniofac J.* 2004, 41(4), 375-80.

[72] Hsieh, CH; Ko, EW; Chen, PK; Huang, CS. The effect of gingivoperiosteoplasty on facial growth in patients with complete unilateral cleft lip and palate. *Cleft Palate Craniofac J.* 2010, 47(5), 439-46.

In: Craniofacial Disorders
Editors: Sh.B. Turner and J. E. Miller
ISBN: 978-1-62618-923-2

Chapter 3

Hearing Loss in Children with Craniofacial Anomalies

Rachael Beswick[1,2] and Carlie Driscoll[2]
[1]Health Services Support Agency, Queensland Health, Queensland Government, Queensland, Australia
[2]Hearing Research Unit for Children, School of Health and Rehabilitation Sciences, The University of Queensland, Brisbane, Queensland, Australia

Abstract

It is well documented that children with craniofacial anomalies are at a greater risk of hearing impairment and middle ear pathology than children without craniofacial anomalies. Most research has focused on the hearing loss present in this population during the first few months of life. However, recent research has indicated that even if these infants pass newborn hearing screening at birth, on-going monitoring of hearing throughout childhood is required due to their increased risk of developing a postnatal hearing loss. A recent study completed by the authors revealed that children with craniofacial anomalies are 2.6 times more likely to develop a postnatal hearing loss than children without craniofacial anomalies. This chapter will begin with an overview of the type and degree of hearing impairment found in children with craniofacial anomalies and the implications of a hearing loss on the child's ability to develop speech and language. It will then explore methods to detect hearing loss from birth through childhood, with an emphasis on on-going monitoring. This chapter will conclude by addressing management options for hearing loss, which will help optimize the child's developmental outcomes.

Introduction

It is well known that there is a strong association between craniofacial anomalies and hearing loss, with approximately 50% of infants with craniofacial anomalies demonstrating some form of hearing loss (Cone-Wesson et al., 2000; de Jong et al., 2010; Hayes, 1994; JCIH, 2007). However, the diagnosis and management of hearing loss can often be delayed while other more life threatening issues, such as ensuring safe airways, are attended to (Swibel Rosenthal, Caballero, & Drake, 2012). Craniofacial anomalies can occur in isolation (e.g., cleft palate), can be associated with a syndrome (e.g., Apert, Pierre Robin), or can occur as a result of a disorder (e.g., metabolic, hematologic). Given the range of complex medical issues in children with craniofacial anomalies, a multidisciplinary holistic approach to management is needed. The role of the audiologist and otolaryngologist is to ensure that optimal hearing is achieved to support the development of speech and language. Although newborn hearing screening can assist with the detection of hearing loss in children at birth, on-going monitoring of hearing throughout childhood is required due to the increased risk of developing a postnatal hearing loss in children with craniofacial anomalies. If a hearing loss is detected in a timely manner, appropriate intervention can occur, which will optimise the child's developmental outcomes.

Incidence, Characteristics, and Underlying Pathology of Hearing Loss in Children with Craniofacial Anomalies

The type and degree of hearing loss present in children with craniofacial anomalies varies significantly and is dependent on the anatomical structures involved and the underlying pathology of the hearing loss. The majority of infants with craniofacial anomalies who present with a hearing loss exhibit a conductive hearing loss, however, sensorineural and mixed hearing losses are also found in these children. In effect, the presence of middle ear pathology may significantly delay the diagnosis of an underlying sensorineural hearing loss as the loss may be considered to be conductive only (Bergstrom, 1978). Previous research by Hayes (1994) and, more recently, de Jong (2010) found of infants with craniofacial anomalies and hearing loss, the hearing loss was conductive in nature in more than 90% of cases. The degree of hearing loss is also dependant on the structures involved and whether there are any other complicating factors. As expected, children with more significant craniofacial anomalies, such as bilateral atresia, will present with a more significant hearing loss, whereas infants with isolated ear anomalies are more likely to present with a mild loss (de Jong, et al., 2010; Hayes, 1994). The following section outlines the most common craniofacial anomalies and their associated hearing losses.

Minor Craniofacial Anomalies (Including Preauricular Skin Tags and/or Ear Pits)

Minor craniofacial anomalies, such as preauricular skin tags and ear pits, are present in approximately 0.3% to 5% of the population (Roth et al., 2008). Although most commonly occurring in isolation, minor craniofacial anomalies may be associated with other major craniofacial anomalies, such as ear canal malformations, or may be associated with genetic syndromes, such as branchio-oto-renal (BOR) syndrome (Fraser, Sproule, & Halal, 1980; Roth, et al., 2008). The incidence of permanent congenital hearing loss reported across studies in this population ranges from 0% to 17% (Firat et al., 2008; Francois, Wiener-Vacher, Falala, & Narcy, 1995; Kankkunen & Thiringer, 1987; Kugelman et al., 1997), with hearing loss more common in children with skin tags than ear pits (Roth, et al., 2008). In fact, some studies have indicated that it is not necessary to investigate hearing in children with preauricular skin tags and ear pits if they occur in isolation (Firat, et al., 2008; Francois, et al., 1995). Preauricular skin tags and ear pits develop during embryogenesis and are caused by defective fusion of the six hillocks that form the auricle (Scheinfeld, Silverberg, Weinberg, & Nozad, 2004; Tan, Constantinides, & Mitchell, 2005). As such, it is expected that the hearing loss would be conductive, rather than sensorineural. Roth et al. (2008) completed a large study on infants with preauricular skin tags and ear pits, with and without additional risk factors. Of the 587 infants with preauricular skin tags and ear pits and no additional risk factors, two infants (0.34%) were diagnosed with a permanent hearing loss.

Both infants presented with a right skin tag and both infants were diagnosed with a bilateral moderate hearing loss (one infant had a bilateral sensorineural hearing loss and the other had a bilateral permanent conductive hearing loss). For those infants with preauricular skin tags and ear pits and additional risk factors, the prevalence of hearing loss was much higher with 13 of 50 infants (26.0%) diagnosed with a permanent hearing loss. Further, for the infants with preauricular skin tags and ear pits and a syndrome and/or another major auricular deformity, the prevalence of hearing loss was 91%. Hearing loss was sensorineural, conductive or mixed, unilateral or bilateral, and ranged from mild to profound (Roth et al., 2008).

Cleft Lip and/or Palate

The incidence of oral clefts ranges from 0.5 to 2 per 1000 live births (Laasonen et al., 2004) and they are often present in children with a syndrome. For children with a syndrome, a cleft palate occurs in isolation approximately 55% of the time and a cleft lip and palate or cleft lip alone occurs approximately 14% of the time (Coleman & Sykes, 2001). Hearing losses in children with a cleft lip and/or palate are due to a number of anatomical abnormalities within the Eustachian tube, skull base, and pharyngeal and nasopharyngeal anatomy (Sheahan, Miller, Sheahan, Earley, & Blayney, 2003; Tuncbilek, Ozgur, & Belgin, 2003). These abnormalities result in chronic conductive hearing loss from middle ear effusions, with otitis media almost universal in this population (Bergstrom, 1978; Chen, Messner, & Curtin, 2008; Hayes, 1994; Luthra, Singh, Nagarkar, & Mahajan, 2009; Sheahan, et al., 2003). There is also a risk of sensorineural hearing loss in children with a cleft lip and/or palate due to the ossicular malformations that occur in many syndromes (Chen, et al., 2008).

The incidence of permanent hearing loss in children with a cleft lip and/or palate varies from 0% to 90% across studies, with 50% of children developing a hearing loss being the reported average (Rohrich & Byrd, 1990). Hayes (1994), who presented one of the primary studies on hearing loss in children with craniofacial anomalies, found that approximately 50% to 60% of children with a cleft lip and/or palate had a permanent hearing loss, and the loss occurred even in instances where the abnormality was in isolation. This report also highlighted that even if a permanent hearing loss was not present, otitis media was widespread in this population (Hayes, 1994).

Microtia and Atresia

Microtia refers to abnormal development of the external ear. The ear deformity that results from microtia ranges from only minor irregularities with auricle landmarks, to the complete absence of the ear (Kellcy & Scholes, 2007). Atresia, which is often associated with microtia, refers to the absence of the external auditory ear canal (Kelley & Scholes, 2007). The incidence of microtia and atresia is variable, with some reports citing one in every 10,000 to 20,000 live births (Kelley & Scholes, 2007), whereas others report one in every 6000 live births (Aase & Tegtmeier, 1977). This variability in incidence is due to the higher rates of microtia and atresia for some populations, specifically the Japanese, Hispanic, and Native Americans (Kelley & Scholes, 2007; Nelson & Berry, 1984). Microtia and atresia often present unilaterally, and are more common in males than females (Kelley & Scholes, 2007).

There are a number of underlying pathological mechanisms for microtia and atresia including abnormal branchial arch development (usually for unilateral microtia), genetic abnormalities (e.g., a single gene deletion in a syndrome), or vascular insults (Kelley & Scholes, 2007). As expected, children with microtia and atresia mostly present with conductive hearing loss, however, there may also be a sensorineural component, particularly if occurring with a syndrome (Roth, et al., 2008). Hayes (1994) reported 100% of children with bilateral atresia presented with a moderate or greater hearing loss, with most infants presenting with a conductive hearing loss. Bergstrom (1978), however, highlighted that the hearing loss may also occur in the contralateral normal-appearing ear.

Syndromes

The relationship between craniofacial anomalies and syndromes is well known with over 400 syndromes identified that have craniofacial anomalies as a characteristic of the syndrome (Lunardi et al., 2012). In fact, the presence of a craniofacial anomaly is often the precursor to numerous genetic and medical tests, which eventually lead to the correct diagnosis of the syndrome (Lunardi, et al., 2012). Syndromes that are highly associated with hearing loss include, but are not limited to: Down syndrome (hearing loss present in 38-78%) (Shott, Joseph, & Heithaus, 2001), Apert syndrome (72%) (de Jong, et al., 2010), Muenke syndrome (67%) (de Jong, et al., 2010), DiGeorge syndrome (55%) (Swibel Rosenthal, et al., 2012), Crouzon/Pfeiffer syndrome (50%) (de

Jong, et al., 2010), Treacher-Collins syndrome (approximately 50%) (Marres, 2002), and Velocardiofacial syndrome (conductive hearing loss present in up to 90%) (McDonald-McGinn et al., 1999).

Most craniofacial anomalies associated with a syndrome are due to developmental abnormalities associated with the first and second branchial arch which can result in malformations in the skeletal, mandibular, muscular, and nervous structures of the face (Johnson, Moonis, Green, Carmody, & Burbank, 2011). For example, hearing loss in children with Down syndrome can be caused by a number of factors including: an abnormal Eustachian tube; malformed auricle, external auditory canals, or middle ear cavity; adenoidal hypertrophy; constriction of the nasopharynx; and mastoid hyperplasia (Balkany, Downs, Jafek, & Krajicek, 1979; Kanamori, Witter, Brown, & Williams-Smith, 2000; Mazzoni, Ackley, & Nash, 1994). For children with DiGeorge syndrome, their ears can be posteriorly rotated, have a simple helix, and/or microtia and anotia may be present (Arndt et al., 2010).

As with all craniofacial anomalies, the type and degree of hearing loss in children with a syndrome varies depending on the structures involved. For some syndromes, the hearing loss is predominately conductive, as in the case of Apert syndrome (up to 80%) (Zhou, Schwartz, & Gopen, 2009). However, other syndromes, such as Crouzon syndrome or DiGeorge syndrome, also present with sensorineural hearing loss (Arndt, et al., 2010; Orvidas, Fabry, Diacova, & McDonald, 1999). When the craniofacial abnormality occurs in conjunction with a syndrome, the management of the hearing loss is often complicated by the other medical issues involved with these children. For a more comprehensive review of craniofacial syndromes and their associated hearing losses, please refer to the review article by Swibel Rosenthal and colleagues (2012) or Lunardi et al. (2012).

Effect of Hearing Loss on Speech and Language Development

The normal processes of speech and language development in children born with craniofacial anomalies may be disrupted by a variety of etiologies including, but not limited to, abnormalities of the oral structures and vocal tract, neuromuscular deficits, hearing loss, and co-morbid disorders such as intellectual impairment. Such children should be referred to appropriate intervention agencies as soon as the primary diagnosis is made (Casper, 1985).

Direct intervention in the form of speech-language therapy and audiological re/habilitation may be required. However, services aimed at promoting speech and language development should not be restricted to speech-language pathologists and audiologists. Achievement of this aim can only be claimed via the close interdisciplinary cooperation of surgeons, orthodontists, ophthalmologists, pediatricians, neurologists, radiologists, psychologists, and social workers. The role of family-centred practice also cannot be understated.

Specifically, children with craniofacial anomalies may experience difficulties in resonance of voice, articulation, and speech intelligibility (Witzel, 1990). Any oro-facial defects have the potential to alter speech production and vocal quality. In particular, defects affecting the size and shape of the tongue, mouth, jaws, and palate can change resonance of the voice (Casper, 1985). Hypernasality, or an excessive 'nasal' vocal quality, is commonly seen in children with cleft palates (Casper, 1985). Hyponasality, or a 'head-cold' vocal quality, may be seen in children with craniofacial anomalies where the passage of air and sound through the nasal tract is blocked (Casper, 1985). The ability to produce the correct pitch and loudness of voice may also be compromised, as may the ability to produce speech sounds that require velopharyngeal competence (Shipley & McAfee, 1998). Fluency disorders can occur in some children, such as those with Fetal Alcohol syndrome. Speech sound production errors (e.g., articulation disorders) may occur as a result of structural abnormalities within the oral cavity as well as oro-motor dysfunction (Casper, 1985), although they may also develop from compensatory strategies. Speech intelligibility, or the suitability of the produced speech to the listener, may be adversely impacted by the quality of voice and articulation of sounds. Phonological processing, the skills involved in producing sound patterns, is also thought to be affected (Witzel, 1990).

Finally, language and literacy development may be comprised as well. Children with some craniofacial anomalies, such as cleft palate for example, may demonstrate a lower quantity and quality of expressive language than their peers. Receptive language deficits are also plausible, in the event of co-morbid hearing loss or intellectual impairment.

Notably, hearing loss has the potential to create or grossly exacerbate many of the speech and language disorders evident in children with craniofacial anomalies. Therefore, full diagnostic hearing assessment at the earliest possible age, followed by regular monitoring of hearing status for the occurrence of late-onset, progressive, or middle ear conditions is crucial. Only in this way may children with craniofacial anomalies have access to the benefits of appropriate early intervention.

Detection of Hearing Loss in Children with Craniofacial Anomalies

For decades, professionals have been aware of the association between craniofacial anomalies and hearing loss and, therefore, have considered it essential to monitor hearing in children with these abnormalities. In fact, 'craniofacial anomalies' (defects of the ear, nose, or throat) was listed as a risk factor for hearing loss on the 1973 Joint Committee on Infant Hearing (JCIH) Position Statement, which recommended that any child with these abnormalities receive regular hearing evaluations (JCIH, 1973). The JCIH is an American-based committee that was established in 1969 due to a growing concern regarding hearing screening procedures in infants. Although the JCIH is based in the United States of America, the Position Statements and recommendations released by this committee are used as a foundation for many hearing programs around the world (e.g., the United Kingdom hearing screening program, Australian hearing screening programs) (Beswick, Driscoll, Kei, & Glennon, 2012b; Russ et al., 2003; Watkin, Hasan, Baldwin, & Ahmed, 2005).

To detect congenital hearing loss, Universal Newborn Hearing Screening (UNHS) programs have been developed and implemented in many parts of the western world. UNHS programs aim to screen all infants at birth, not just those with risk factors. The screen is usually performed within the first 6 to 48 hours of the infant's life. Typically, the infant is screened in hospital before discharge. However, if this is not possible (e.g., due to home birth), the infant should be screened within the first month of life (Centers for Disease Control and Prevention, 2010; JCIH, 2007). Protocols for hearing screening vary significantly around the world. However, the infant is usually screened by a nurse, trained individual, or technician, using either Otoacoustic Emissions testing (OAEs) or an automated Auditory Brainstem Response (aABR) (Diefendorf, 2002; JCIH, 2007). These assessment tools are used as they are non-invasive physiological measures of the auditory system which allow automatic delivery and simple result interpretation (Diefendorf, 2002; Norton et al., 2000). If the infant obtains a 'refer' result on the initial screen, a second screen is required. This screen is also performed within the first month of life and is completed to help minimise over-referral to audiology and medical services (JCIH, 2007). Most infants have passed the screen by the second attempt (JCIH, 2007). However, if a 'refer' result is once again obtained, referral to an audiologist is required. At the audiology appointment, a variety of diagnostic assessments will be performed to help determine the type, degree, and configuration of the hearing loss.

For the general population, it is expected that UNHS programs will detect approximately 1 to 2 per 1000 infants with a bilateral hearing loss of moderate degree or greater (Fortnum, Summerfield, Marshall, Davis, & Bamford, 2001; Watkin & Baldwin, 2011). However, for children with craniofacial anomalies, given the anatomical structures involved and the almost universality of conductive hearing loss, significantly more children refer on the newborn hearing screen and are subsequently identified with a permanent hearing loss, even when the craniofacial anomaly is only minor (Chen, et al., 2008; Hayes, 1994). For example, for children with preauricular skin tags and ear pits, the prevalence of hearing loss increases to 8 per 1000 children (Roth et al., 2008). In addition, Chen et al. (2008) found of 30 infants with craniofacial abnormalities who failed newborn hearing screening and were referred to audiology, 13 infants were identified with a permanent hearing loss (12 with a sensorineural hearing loss, 1 with a mixed hearing loss).

When children with a craniofacial anomaly are examined within a population study, similar results are presented. In England, of the 169,487 infants screened, 169 infants were detected with a bilateral hearing loss of a moderate degree or greater. Of these 169, 11 infants presented with craniofacial anomalies (6.5% of all hearing losses had craniofacial anomalies) (Uus & Bamford, 2006). Similarly, in New York where 43,311 infants were screened, 85 infants were detected with a hearing loss. Of these 85 infants, 11 infants (12.9%) presented with craniofacial anomalies (Dalzell et al., 2000). In both programs, a combination of OAE and aABR screening equipment was used.

Despite the above findings, there are some concerns with using a newborn hearing screen to detect hearing loss in all children with craniofacial anomalies. Newborn hearing screening programs will not detect hearing loss that develops postnatally, or children with mild or minimal losses, or those with high-frequency losses (Bamford et al., 2007). Given the type of hearing loss present in children with craniofacial anomalies, OAE screening may be more effective in detecting hearing loss than aABR screening due to the part of the auditory pathway being tested (outer hair cells of the cochlea versus the auditory nerve, respectively) and its greater susceptibility to middle and outer ear issues (Leigh, Schmulian-Taljaard, & Poulakis, 2010). Although more effective for this population, fewer programs opt for an OAE only screening protocol, given the higher referral rates and inability to detect auditory neuropathy (Leigh, et al., 2010). Further, some manufacturers have provided statements (e.g. Natus who have released the Algo aABR) specifying that the equipment is not intended to be used on newborns with abnormal outer, middle, and inner ear anomalies, obvious deformities of the head and neck, and infants with neurological conditions (Natus Medical, n.d.). However, in some programs (e.g., Queensland, Australia), all babies are screened with this equipment unless there is a major craniofacial abnormality including unilateral or bilateral microtia or atresia, or if the treating clinician overrides the screen and refers directly to audiology. This protocol, to screen all eligible babies, is adopted to ensure the universality of the newborn hearing screening program.

Detection of Hearing Loss in Children with Craniofacial Anomalies beyond the Newborn Period

Despite the implementation of UNHS programs, continued monitoring of hearing in children with craniofacial anomalies is required, due to their increased risk of developing a postnatal hearing loss. As mentioned above, not all hearing losses present in these children can be detected by the newborn hearing screen. Only recently have publications begun to emerge on children who have passed newborn hearing screening but later have developed a postnatal hearing loss. Currently, targeted surveillance of at-risk infants has been recommended by the JCIH to detect hearing loss *post* newborn hearing screening (JCIH, 2007). During the birth screen or early well-child visits, parents are interviewed by a nurse/midwife regarding any potential risk factors that may lead to a postnatal hearing loss. Risk factors vary between screening programs but may include factors such as a family history of hearing loss, syndromes associated with hearing loss, neurodegenerative disorders, cytomegalovirus, extracorporeal membrane oxygenation, culture positive infections, chemotherapy, and craniofacial anomalies. If a child presents with one or more of these risk factors, then they are referred to a targeted surveillance program for hearing monitoring. The JCIH (2007) recommend at least one audiology assessment by 24 to 36 months of age, with children with some risk factors, such as CMV, requiring more frequent assessments.

The current authors completed an extensive analysis of a targeted surveillance program in Queensland, Australia. This targeted surveillance program uses a risk factor registry that was based on the JCIH (2000) risk factor list. In Queensland, the risk factor 'craniofacial anomalies' includes, but is not limited to: facial anomalies (mouth, eyes, chin), cleft palate, facial paralysis, craniosyntosis, microcephaly, hydrocephaly, and neck webbing. During the study period of 5 years, 4 months, 7,320 (of 261,328 screened) children were referred for targeted surveillance, of which 588 children were referred for monitoring due to the risk factor of craniofacial anomalies. Of these 588 children, 10 (1.7%) developed a postnatal hearing loss. Of all the risk factors, the risk factor of craniofacial anomalies presented with the second highest yield of all the children with postnatal hearing, falling behind syndrome (Beswick, et al., 2012b). These findings are similar to those reported by Chen et al. (2008) who reported that of the 74 children with craniofacial anomalies who passed newborn hearing screening, two children (2.7%) were later diagnosed with permanent hearing loss. In the Queensland study, further examination using logistic regression analysis was completed on a proportion of the cohort , revealing that children with craniofacial anomalies were 2.6 times more likely to develop a postnatal hearing loss than children without craniofacial anomalies (Beswick, Driscoll, Kei, Khan, & Glennon, 2013).

Several alternate methods to targeted surveillance have also been reported in the literature to detect postnatal hearing loss. These methods include active measures, such as child care and school screening programs, and responsive measures, such as professional/parental concerns regarding hearing and delays in speech and language development (Beswick, et al., 2012b; Dedhia, Kitsko, Sabo, & Chi, 2013; Lü et al., 2011; Watkin & Baldwin, 2011). In all studies, it was only possible to detect postnatal hearing loss through use of multiple post-neonatal care pathways as, unlike UNHS programs, there currently exists no universal approach for capturing postnatal hearing loss. For example, Watkin and Baldwin (2011) utilised a health visitor distraction test, parental and professional concerns and a school entry hearing test to identify children with postnatal hearing loss. In their study, a craniofacial anomaly was reported as a major risk factor present in children with mild, late onset, hearing loss. Similarly, Lu et al. (2011) utilised a preschool screening program to detect hearing loss post newborn hearing screening. Of the 445 children assessed at preschool, 16 presented with a postnatal hearing loss, and one had a craniofacial anomaly. This child was detected with a unilateral moderate hearing loss.

The above studies stress the importance of monitoring hearing in children with craniofacial anomalies due to their increased risk of developing a postnatal hearing loss. In fact, the importance of monitoring hearing in this population has been raised in a recent publication that provided recommendations for hearing monitoring programs. This recommendation stated that 'Children with craniofacial anomalies (excluding ear pits and skin tags) as a risk factor should have their hearing monitored throughout early childhood'(Beswick, Driscoll, Kei, & Glennon, 2012a, p. 63).

Management of Hearing Loss

Management of hearing loss in children with craniofacial anomalies is variable given the range of hearing losses found in these children. For optimal outcomes, a multidisciplinary approach is recommended and should include an audiologist, otolaryngologist, general practitioner, pediatrician, speech-language pathologist, and teacher of the deaf. Depending on the type of hearing loss, management can be either conservative or aggressive, with significant debate surrounding the best approach to take (Luthra, et al., 2009). For example, for middle ear pathology, some otolaryngologists prefer a 'watch and wait' approach and only insert grommets for persistent, bilateral otitis media. However, other professionals may be more aggressive in their management and insert grommets early to prevent hearing loss (Elloy & Marshall, 2011; Luthra, et al., 2009). For some children (e.g., in Down syndrome), grommet insertion may not be viable given the ear canals are too narrow. In these cases, hearing aids may be more appropriate. Even in cases where surgery is an option, amplification may be utilised until the surgery is performed (Hayes, 1994).

Over the last decade, hearing aid technology has advanced significantly with multiple brands and models now available. Typically, digital behind-the-ear hearing aids are the standard amplification prescribed by audiologists for children with a hearing loss. Behind-the-ear hearing aids are light-weight and are small enough for pediatric ears, even for those with a misshaped pinna. However, for children with chronic otitis media, or microtia and/or atresia, a bone conductor hearing aid may be more suitable. A bone conductor hearing aid sits on the mastoid bone behind the ear and directly stimulates the cochlea. A bone anchored hearing aid (BAHA) is another option for children where conventional hearing aids are not appropriate. A BAHA is an implantable device that is inserted into the calvarial bone behind the ear and is usually used for children over 4 years old (Elloy & Marshall, 2011). Another implantable device is a cochlear implant. Cochlear implants can be fitted unilaterally or bilaterally and are indicated in children with a bilateral severe to profound hearing loss who are not receiving adequate auditory input from hearing aids. Children who require a cochlear implant are assessed by a multidisciplinary cochlear implant team to ensure they are suitable for this type of amplification. Regardless of the intervention, be it surgery or amplification, it is recommended that it occurs as early as possible, to help optimise speech, language, and hearing outcomes for the child.

Conclusion

Children with craniofacial anomalies are at a greater risk of hearing impairment and middle ear pathology than children without craniofacial anomalies. If undetected, hearing impairment can have a major impact on the child's speech and language development. Newborn hearing screening programs have been implemented to help detect hearing loss in children from birth. However, given the increased risk of developing a hearing loss in children with craniofacial anomalies, continued monitoring of hearing throughout childhood is recommended. If a hearing loss is detected, timely surgical intervention or amplification will help minimise the impact upon the child's development.

References

Aase, J. M., & Tegtmeier, R. E. (1977). Microtia in New Mexico: Evidence for multifactorial causation. *Birth Defects Original Article Series, 13*, 113-116.

Arndt, S., Laszig, R., Beck, R., Schild, C., Maier, W., Birkenhager, R., Aschendorff, A. (2010). Spectrum of hearing disorders and their management in children with CHARGE syndrome. *Otology and Neurotology, 31*(1), 67-73.

Balkany, T. J., Downs, M. P., Jafek, B. W., & Krajicek, M. J. (1979). Hearing loss in Down's syndrome. A treatable handicap more common than generally recognized. *Clinical Pediatrics, 18*(2), 116-118.

Bamford, J., Fortnum, H., Bristow, K., Smith, J., Vamvakas, G., Davies, L., Hind, S. (2007). Current practice, accuracy, effectiveness and cost-effectiveness of the school entry hearing screen. *Health Technology Assessment, 11*(32), 1-168.

Bergstrom, L. (1978). Congenital and acquired deafness in clefting and craniofacial syndromes. *Cleft Palate Journal, 15*(3), 254-261.

Beswick, R., Driscoll, C., Kei, J., & Glennon, S. (2012a). Recommendations for monitoring hearing in children using a risk factor registry. *Journal of Hearing Science, 2*(2), 57-65.

Beswick, R., Driscoll, C., Kei, J., & Glennon, S. (2012b). Targeted surveillance for postnatal hearing loss: A program evaluation. *International Journal of Pediatric Otorhinolaryngology, 78*, 1046-1056.

Beswick, R., Driscoll, C., Kei, J., Khan, A., & Glennon, S. (2013). Which risk factors predict postnatal hearing loss in children? *Journal of the American Academy of Audiology, 24*(3), 205-213.

Casper, J. (1985). Disorders of speech and voice. *Pediatric Annals, 14*(3), 220-221.

Centers for Disease Control and Prevention. (2010). *Identifying infants with hearing loss- United States: 1999-2007.* Retrieved April 23, 2012, from http://www.cdc.gov/mmwr/preview/mmwrhtml/mm5908a2.htm

Chen, J. L., Messner, A. H., & Curtin, G. (2008). Newborn hearing screening in infants with cleft palates. *Otology and Neurotology, 29*(6), 812-815.

Coleman, J. R., & Sykes, J. M. (2001). The embryology, classification, epidemiology, and genetics of facial clefting. *Facial Plastic Surgery Clinics of North America, 9*(1), 1-13.

Cone-Wesson, B., Vohr, B. R., Sininger, Y. S., Widen, J. E., Folsom, R. C., Gorga, M. P., & Norton, S. J. (2000). Identification of neonatal hearing impairment: Infants with hearing loss. *Ear and Hearing, 21*(5), 488-507.

Dalzell, L., Orlando, M., MacDonald, M., Berg, A., Bradley, M., Cacace, A., Prieve, B. (2000). The New York State universal newborn hearing screening demonstration project: Ages of hearing loss identification, hearing aid fitting, and enrollment in early intervention. *Ear and Hearing, 21*(2), 118-130.

de Jong, T., Bannink, N., Bredero-Boelhouwer, H. H., van Veelen, M. L. C., Bartels, M. C., Hoeve, L. J., ... Mathijssen, I. M. J. (2010). Long-term functional outcome in 167 patients with syndromic craniosynostosis: Defining a syndrome-specific risk profile. *Journal of Plastic, Reconstructive & Aesthetic Surgery, 63*(10), 1635-1641.

Dedhia, K., Kitsko, D., Sabo, D., & Chi, D. (2013). Children with sensorineural hearing loss after passing the newborn hearing screen. *JAMA Otolaryngology - Head and Neck Surgery, 139*(2), 119-123.

Diefendorf, A. O. (2002). Detection and assessment of hearing loss in infants and children. In J. Katz (Ed.), *Handbook of clinical audiology* (5th ed., pp. 469-480). Philadelphia: Lippincott Williams & Wilkins.

Elloy, M. D., & Marshall, A. H. (2011). The management of hearing loss. *Paediatrics and Child Health, 22*(1), 13-18.

Firat, Y., Sireci, S., Yakinci, C., Akarcay, M., Karakas, H. M., Firat, A. K., . . . Selimoglu, E. (2008). Isolated preauricular pits and tags: Is it necessary to investigate renal abnormalities and hearing impairment? *European Archives of Oto-Rhino-Laryngology, 265*(9), 1057-1060.

Fortnum, H., Summerfield, A., Marshall, D., Davis, A., & Bamford, J. (2001). Prevalence of permanent childhood hearing impairment in the United Kingdom and implications for universal neonatal hearing screening: Questionnaire based ascertainment study. *British Medical Journal (Clinical Research Edition), 323*(7312), 536-540.

Francois, M., Wiener-Vacher, S. R., Falala, M., & Narcy, P. (1995). Audiological assessment of infants and children with preauricular tags. *Audiology, 34*(1), 1-5.

Fraser, F. C., Sproule, J. R., & Halal, F. (1980). Frequency of the branchio-oto-renal (BOR) syndrome in children with profound hearing loss. *American Journal of Medical Genetics, 7*(3), 341-349.

Hayes, D. (1994). Hearing loss in infants with craniofacial anomalies. *Otolaryngology - Head and Neck Surgery, 110*(1), 39-45.

JCIH. (1973). Screening for infant hearing Retrieved October 7, 2007, from http://www.jcih.org/posstatemts.htm

JCIH. (2000). Year 2000 position statement: Principles and guidelines for early hearing detection and intervention programs. *Pediatrics, 106*(4), 798-817.

JCIH. (2007). Year 2007 position statement: Principles and guidelines for early hearing detection and intervention programs. *Pediatrics, 120*(4), 898-921.

Johnson, J. M., Moonis, G., Green, G. E., Carmody, R., & Burbank, H. N. (2011). Syndromes of the first and second branchial arches, part 1: Embryology and characteristic defects. *American Journal of Neuroradiology, 32*(1), 14-19.

Kanamori, G., Witter, M., Brown, J., & Williams-Smith, L. (2000). Otolaryngologic manifestations of Down syndrome. *Otolaryngologic Clinics of North America, 33*(6), 1285-1292.

Kankkunen, A., & Thiringer, K. (1987). Hearing impairment in connection with preauricular tags. *Acta Paediatrica Scandinavica, 76*(1), 143-146.

Kelley, P. E., & Scholes, M. A. (2007). Microtia and congenital aural atresia. *Otolaryngologic Clinics of North America, 40*(1), 61-80.

Kugelman, A., Hadad, B., Ben-David, J., Podoshin, L., Borochowitz, Z., & Bader, D. (1997). Preauricular tags and pits in the newborn: The role of hearing tests. *Acta Paediatrica, 86*(2), 170-172.

Laasonen, M., Haapanen, M., Maenpaa, P., Pulkkinen, J., Ranta, R., & Virsu, V. (2004). Visual, auditory, and tactile temporal processing in children with oral clefts. *Journal of Craniofacial Surgery, 15*(3), 510-518.

Leigh, G., Schmulian-Taljaard, D., & Poulakis, Z. (2010). Newborn hearing Screening. In C. J. Driscoll & B. McPherson (Eds.), *Newborn screening systems: The complete perspective* (pp. 95-115). San Diego: Plural Publishing.

Lü, J., Huang, Z., Yang, T., Li, Y., Mei, L., Xiang, M., . . . Wu, H. (2011). Screening for delayed-onset hearing loss in preschool children who previously passed the newborn hearing screening. *International Journal of Pediatric Otorhinolaryngology, 75*(8), 1045-1049.

Lunardi, S., Forli, F., Michelucci, A., Liumbruno, A., Baldinotti, F., Fogli, A., Simi, P. (2012). *Genetic hearing loss associated with craniofacial abnormalities.*

Luthra, S., Singh, S., Nagarkar, A. N., & Mahajan, J. K. (2009). The role of audiological diagnostics in children with cleft lip & palate (CLP). *International Journal of Pediatric Otorhinolaryngology, 73*(10), 1365-1367.

Marres, H. A. M. (2002). Hearing loss in the Treacher-Collins syndrome. *Advances in Oto-Rhino-Laryngology, 61*, 209-215.

Mazzoni, D. S., Ackley, R. S., & Nash, D. J. (1994). Abnormal pinna type and hearing loss correlations in Down's syndrome. *Journal of Intellectual Disability Research, 38*(Pt 6), 549-560.

McDonald-McGinn, D. M., Kirschner, R., Goldmuntz, E., Sullivan, K., Eicher, P., Gerdes, M., . . . Zackai, E. H. (1999). The Philadelphia story: The 22q11.2 deletion- report on 250 patients. *Genetic Counseling, 10*(1), 11-24.

Natus Medical. (n.d.). Quick reference guide for using the ALGO 3i newborn hearing screener. Retrieved from http://www.natus.com.

Nelson, S. M., & Berry, R. I. (1984). Ear disease and hearing loss among Navajo children: A mass survey. *Laryngoscope, 94*(3), 316-323.

Norton, S. J., Gorga, M. P., Widen, J. E., Folsom, R. C., Sininger, Y., Cone-Wesson, B., Fletcher, K. (2000). Identification of neonatal hearing impairment: Evaluation of transient evoked otoacoustic emission, distortion product otoacoustic emission, and auditory brain stem response test performance. *Ear and Hearing, 21*(5), 508-528.

Orvidas, L. J., Fabry, L. B., Diacova, S., & McDonald, T. J. (1999). Hearing and otopathology in Crouzon syndrome. *Laryngoscope, 109*(9), 1372-1375.

Rohrich, R. J., & Byrd, H. S. (1990). Optimal timing of cleft palate closure. Speech, facial growth, and hearing considerations. *Clinics in Plastic Surgery, 17*(1), 27-36.

Roth, D. A., Hildesheimer, M., Bardenstein, S., Goidel, D., Reichman, B., Maayan-Metzger, A., & Kuint, J. (2008). Preauricular skin tags and ear pits are associated with permanent hearing impairment in newborns. *Pediatrics, 122*(4), e884-890.

Russ, S. A., Poulakis, Z., Barker, M., Wake, M., Rickards, F., Saunders, K., & Oberklaid, F. (2003). Epidemiology of congenital hearing loss in Victoria, Australia. *International Journal of Audiology, 42*(7), 385-390.

Scheinfeld, N. S., Silverberg, N. B., Weinberg, J. M., & Nozad, V. (2004). The preauricular sinus: A review of its clinical presentation, treatment, and associations. *Pediatric Dermatology, 21*(3), 191-196.

Sheahan, P., Miller, I., Sheahan, J. N., Earley, M. J., & Blayney, A. W. (2003). Incidence and outcome of middle ear disease in cleft lip and/or cleft palate. *International Journal of Pediatric Otorhinolaryngology, 67*(7), 785-793.

Shipley, K., & McAfee, J. (1998). *Assessment in speech language pathology: A resource manual* (2nd ed.). San Diego: Singular.

Shott, S. R., Joseph, A., & Heithaus, D. (2001). Hearing loss in children with Down syndrome. *International Journal of Pediatric Otorhinolaryngology, 61*(3), 199-205.

Swibel Rosenthal, L. H., Caballero, N., & Drake, A. F. (2012). Otolaryngologic manifestations of craniofacial syndromes. *Otolaryngologic Clinics of North America, 45*(3), 557-577.

Tan, T., Constantinides, H., & Mitchell, T. E. (2005). The preauricular sinus: A review of its aetiology, clinical presentation and management. *International Journal of Pediatric Otorhinolaryngology, 69*(11), 1469-1474.

Tuncbilek, G., Ozgur, F., & Belgin, E. (2003). Audiologic and tympanometric findings in children with cleft lip and palate. *Cleft Palate-Craniofacial Journal, 40*(3), 304-309.

Uus, K, & Bamford, J. (2006). Effectiveness of population-based newborn hearing screening in England: Ages of interventions and profile of cases. *Pediatrics, 117*(5), e887-893.

Watkin, P. M., & Baldwin, M. (2011). Identifying deafness in early childhood: Requirements after the newborn hearing screen. *Archives of Disease in Childhood, 96*(1), 62-66.

Watkin, P. M., Hasan, J, Baldwin, M., & Ahmed, M. (2005). Neonatal hearing screening: Have we taken the right road? Results from a 10-year targeted screen longitudinally followed up in a single district. *Audiological Medicine, 3*, 175-184.

Witzel, M. A. (1990). Craniofacial anomalies. *Seminars in Speech and Language, 11*(3), 145-156.

Zhou, G., Schwartz, L. T., & Gopen, Q. (2009). Inner ear anomalies and conductive hearing loss in children with Apert syndrome: An overlooked otologic aspect. *Otology and Neurotology, 30*(2), 184-189.

In: Craniofacial Disorders
Editors: Sh.B. Turner and J. E. Miller
ISBN: 978-1-62618-923-2

Chapter 4

Strabismus in Craniosynostosis

Zia Chaudhuri[1*], Fiona Dean[2], Anamika Tandon[3], Ajay Sharma[4] and Joseph L Demer[5]

[1]Strabismus & Neuro-ophthalmology Services, Maulana Azad Medical College & Associated Hospitals, New Delhi, India
[2]University Hospitals of Coventry and Warwickshire, UK
[3]Leicester Royal Infirmary, University of Leicester, Leicester, UK
[4]Pediatric Craniofacial Unit, G B Pant Hospital, Maulana Azad Medical College, New Delhi, India
[5]Division of Pediatric Ophthalmology & Strabismus, Ocular Motility Labs, Jules Stein Eye Institute, University of California, Los Angeles, CA, US

Abstract

Craniosynostosis is caused by abnormal premature fusion of any cranial suture. Because the orbits bridge the face to the cranium, ocular and adnexal abnormalities are common in many forms of craniosynostosis. Phenotypic classifications of craniosynostosis were initially developed, the simplest being non-syndromic craniosynostosis

* Corresponding Author: Dr Zia Chaudhuri. Email: ziachaudhuri1@gmail.com; Phone: +911122759613 (Landline); +919968604385 (Office Mobile); +919810645168 (Mobile).

including scaphocephaly, trigonocephaly, synostotic plagiocephaly, versus syndromic craniosynostosis including the syndromes of Crouzon, Apert, Pfeiffer and Saethre-Chotzen. [1] More recently, phenotypic classifications have been augmented by more mechanistic genetic studies of the fibroblast growth factor receptor genes, FGFR1, FGFR2, FGFR3 and TWIST genes, which have not corresponded well to phenotypic classifications. [2]

Strabismus is observed in 39 to 91 % of patients with syndromic craniosynostosis. [3]. All types of strabismus may be seen. [4-7] Mechanisms of strabismus in such cases are manifold, including heterotopy of muscle pulleys, absent muscles, and muscle force imbalance due to orbital asymmetry. [3, 8-10] Non-syndromic craniosynostosis are also associated with increased incidence of refractive errors and strabismus. [11] Strabismus, ptosis, uncorrected refractive errors, and corneal exposure are all amblyogenic and can cause significant visual disability in children with craniosynostosis. [12, 13]

Diagnosis and treatment of strabismus and amblyopia in patients with craniosynostosis is important towards preservation of functional vision. Imaging provides objective parameters to determine appropriate strabismus surgery. [11]

Improved neuro-surgical treatment in these patients has resulted in increased survival but there is evidence to show that the strabismus may change after such interventions. [14-18] A variety of surgical maneuvers have been described to correct the strabismus. [19-22] A multidisciplinary team comprising plastic surgeons, neurosurgeons, maxillofacial surgeons and ophthalmologists is valuable in preserving vision in these patients.

Introduction

Craniosynostosis denotes premature closure of cranial sutures. While suture closure terminates bony growth perpendicular to the line of the affected suture, growth continues parallel to the suture, causing characteristic skull and facial features. [3, 23-26] At birth, the cranial part of the skull is large compared with the face, which is small due to underdeveloped sinuses. The newborn skull thus has an ovoid shape with a wider posterior part to accommodate the brain. The anterior part of the skull is formed by the frontal bone, which extends posteriorly to the coronal suture, and anteriorly forms the roof of the orbit overhanging the nasal bones. [1, 24, 27] The calvarial bones, arising at widely separated primary ossification centers, spread towards each other and are separated by fibrous or membranous tissues, continuous externally with the periosteum and internally with the endosteal layer of the dura. The sutures are primary growth sites. This expansion is secondary to volumetric expansion of the neural mass comprising the brain, leptomeninges and the cerebro-spinal fluid (CSF).

Although craniosynostosis is variably distributed in different geographical locations globally, on an average it is reported to occur in about 1 in 2500 births [3, 23], with about 80 % patients presenting with non-syndromic craniosynostosis, with 10-14 % genetic transmissibility and 20% with syndromic craniosynostosis. [28] Crouzon's syndrome, Apert's syndrome, Pfeiffer's syndrome, Carpenter's syndrome, Muenke's syndrome and Saethre-Chotzen syndrome comprise some common syndromic craniosynostosis with mostly autosomal dominant inheritance. [27, 28] Commonly used nomenclature for non-syndromic craniosynostosis is based on the description of the resultant skull deformity due to the sutural synostosis. This comprises names like scaphocephaly for a long, narrow skull, which is a result of stenosis along the sagittal suture; brachycephaly, an abnormally wide head due to closure of the bicoronal sutures, unicoronal craniosynostosis resulting in plagiocephaly, metopic suture synostosis resulting in trigonocephaly and stenosis of the coronal and the lambdoid sutures resulting in oxycephaly (Figures 1 to 5). [1, 27]

Timely neuro-surgical management has improved the quality of life of patients with plagiocephaly, so that other abnormalities, like ophthalmic disorders that had hitherto been subordinated to life threatening primary anomalies have gained relative importance. [3, 6, 8, 10-12, 15, 16, 29-39] With the possible exception of cosmetic considerations, the visual system is more impacted by craniosynostosis of all types than any other. [4, 6, 12, 13, 15, 16, 30-32, 34-38, 40-46] Many ophthalmic abnormalities associated with craniosynostosis have been described from different parts of the world. [6, 10-12, 16, 28, 30, 33, 39, 40, 44, 47-53] Structural changes due to craniosynostosis may cause orbital shallowing and pseudoprotosis, orbital excyclorotation leading to hypertelorism, and optic nerve edema and atrophy due to progressive compression. [3, 11-13, 39, 54]

Functional visual abnormalities like amblyopia comprise further morbidities in craniosynostosis, either mediated by strabismus and ocular motility abnormalities, or as consequences of structural abnormalities in the orbit and ocular adnexa in these patients. [30] Neurosurgical improvements have left amblyopia as the most common cause of visual impairment in these children, replacing optic nerve atrophy. [12, 38, 48, 55] Although reversible if treated early, amblyopia due to uncorrected refractive errors, ptosis, and strabismus can result in permanent visual loss. Other causes of visual loss include corneal exposure due to lagophthalmos resulting from pseudo-proptosis, and optic atrophy due to progressive optic nerve compression. [4, 12, 13, 15, 16, 38, 39, 45, 48, 55] Unicoronal craniosynostosis has been associated with anisometropia, especially aniso-astigmatism in both the affected and contralateral sides, with or without strabismus (Figure 5). [6, 18, 31, 35] Suspicion of these subtle anomalies, along with early diagnosis and prompt care of the more obvious complications, improves visual prognosis for these increasingly treatable conditions resulting in a better quality of life.

Trigonocephaly, a rare form of craniosynostosis, caused by metopic suture synostosis, is associated with low degrees of strabismus and astigmatism, but the impact on visual development may be significant if the skull deformities are corrected late. [14, 16, 55] As the incidence of significant refractive error requiring correction and strabismus was found to be higher in patients with metopic suture stenosis as compared to the general population, orthoptic and ophthalmic surveillance is recommended in these patients. [55]

Naso-lacrimal duct obstruction (NLDO) is commonly observed in patients with Apert's syndrome, especially in those having the Ser352Trp and Pro253Arg mutations on the FGRF2 gene. [33]

Clinical Patterns of Strabismus in Craniosynostosis

Strabismus is common in syndromic craniosynostosis with rates varying from 39% to 91% in different studies. [3] Exotropia, especially with the V pattern incomitance, is commonly reported. [3, 7, 11, 18, 28, 36, 47] Conversely, in a large retrospective European study comprising almost 14000 patients with A and V pattern strabismus, it was observed that while patients with A pattern strabismus demonstrated greater etiological prevalence of neurological impairment, neural tube defects and hydrocephalus, patients with V pattern strabismus exhibited greater prevalence of craniosynostosis and other malformations. [56] Esotropia is relatively uncommon as compared to exotropia but has been more commonly observed in Apert's syndrome than in other forms of craniosynostosis. [3, 43, 46]

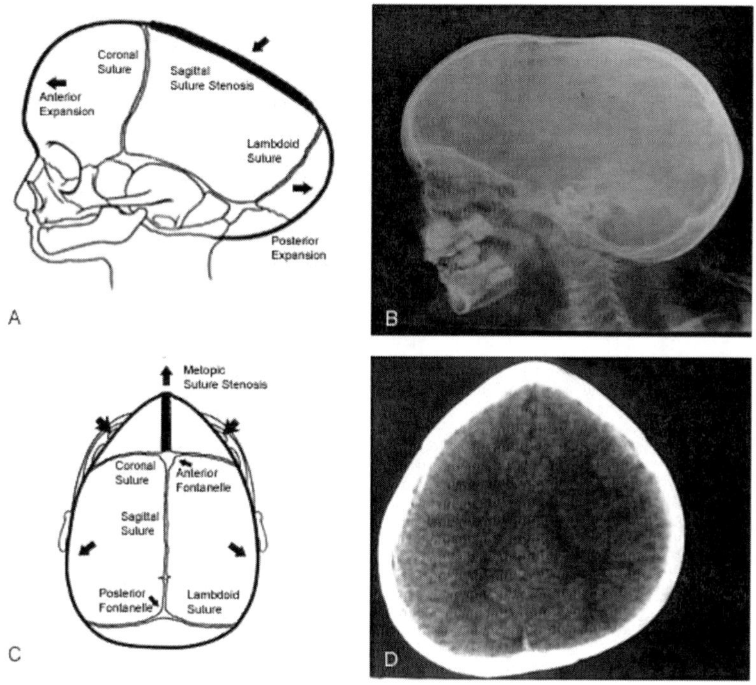

Figure 1. A to D: (A) Diagrammatic representation of the anterior and posterior expansion of the skull due to sagittal suture stenosis, resulting in the classical boat shaped skull characteristic of scaphocephaly; (B) A digital skiagram of a scaphocephalic skull demonstrates the classical copper beaten appearance; (C) Diagrammatic representation of metopic suture stenosis resulting in bilateral posterior expansion and the characteristic triangular trigonocephalic skull; (D) Axial CT scan of a trigonocephalic skull shows the typical triangular shape associated with marked distortion of cerebral matter (Photo courtesy: Professor Ajay Sharma).

Genetics in Strabismus

Genetic mutations on the FGRF2 gene causing specific amino acid substitutions such as Ser352Trp and Pro253Arg have been associated with higher rates of strabismus in Apert's syndrome. [33, 34, 37, 38] There was a trend of more frequent strabismus and amblyopia associated with the FGFR2Ser352Trp mutation and more frequent optic disc pallor with the FGFR2Pro253Arg mutation. [34] Significantly greater prevalence of visual impairment was observed in patients with the FGFR2Ser322Trp mutation as compared to the FGFR2Pro253Arg mutation. [34] Ocular phenotype correlations in patients with the Saethre-Chotzen craniosynostosis and Muenke's craniosynostosis with the TWIST and FRFR3 gene revealed that patients with mutations of the TWIST gene presented with more ophthalmic abnormalities in the form of strabismus, especially vertical mis-alignment, ptosis, NLDO, astigmatism and amblyopia. [32, 40] Strabismus in oblique gaze directions, dissociated vertical deviations, and absence of extraocular muscles, especially the vertical recti, all have been associated with hypertropia in both syndromic and non-syndromic craniosynostosis. [3, 5, 8, 11, 17, 22, 42, 47, 51, 57-59] Imaging studies interpreted through biomechanical modeling have suggested that rectus pulley displacement can cause ipsilateral hypertropia in patients with non-syndromic unicoronal craniosynostosis. [8] In addition, it has been reported that patients with unicoronal craniosynostosis with a mutation of the FGFR2 or FGFR3 gene usually have manifest strabismus, similar to that seen with syndromic craniosynostosis. [6] All types of non-syndromic craniosynostosis have been associated with increased incidence of significant refractive errors and horizontal strabismus, similar to but possibly not to the extent seen in syndromic craniosynostosis. [3, 11] Thus, various mechanisms, genetic, structural and functional, should be considered while determining the causes and manifestations of strabismus in craniosynostosis. Orbital imaging to determine the exact cause of strabismus is recommended for pre-surgical planning in this milieu. [11]

Mechanisms of Strabismus in Craniosynostosis

Mechanisms of strabismus include absence of individual extra-ocular muscles (EOMs), [60-65] sometimes multiple EOMs simultaneously, [58] anomalous and redundant EOMs with decreased molecular functionality, [63, 64] abnormal configurations such as Y-split of the medial rectus (MR) muscle, [19] posterior insertions and sagittalization of rectus and oblique EOMs, [8, 66] abnormal shape and dimensions of the orbit and EOMs with the change in their relation to each other resulting in heterotopy, instability and displacement of connective tissue pulleys of EOMs, [8] abnormal orbital configuration associated with divergent orbital axis or exorbitism and hypertelorism, and sometimes, sensory strabismus. [59] Compression of the globe induced by the bony deformation due to craniosynostosis has also been suggested as a possible cause for the increased incidence of strabismus and astigmatism associated with the condition. [14, 64] EOM dysfunction in unicoronal plagiocephaly has been postulated to be a major cause of abnormal head posture in these patients. The torticollis decreased in all patients in this group upon resolution of strabismus, either subsequent to craniofacial surgery or to strabismus surgery. [17, 42, 59]

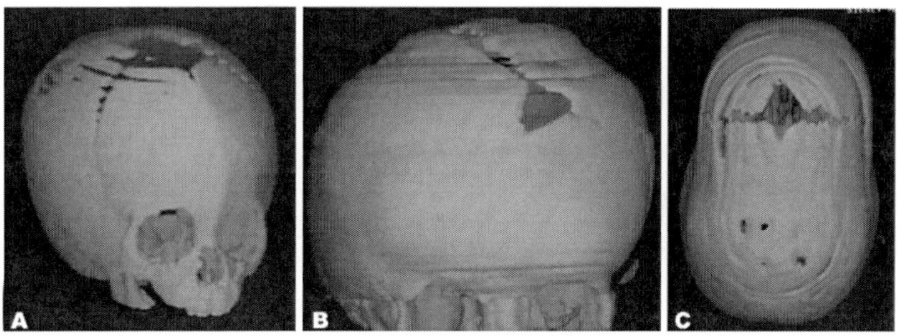

Figure 2. A, B and C: 3D CT reconstruction of synostosis of the metopic suture (A) resulting in trigonocephly, the right sided coronal suture (B) leading to unicoronal plagiocephaly, and sagittal suture (C) leading to scaphocephaly. The abnormal skull shapes are a consequence of expansion perpendicular to the direction of the prematurely closed suture. (Photo courtesy: Professor Ajay Sharma) .

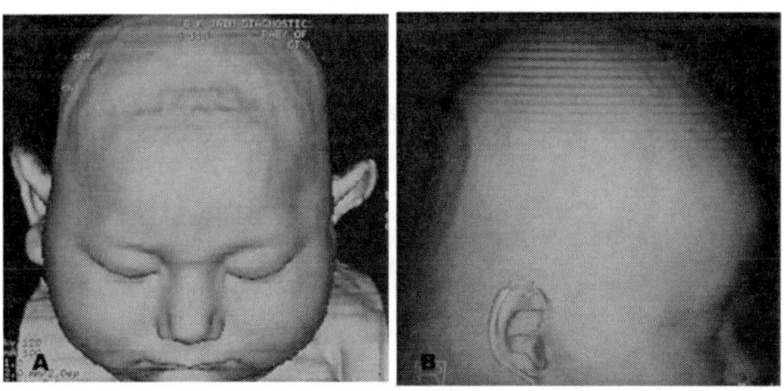

Figure 3. A and B: Coronal and sagittal 3D reconstruction of the appearance of a patient with bicoronal and lambdoid suture stenosis resulting in oxycephaly (Photo courtesy: Professor Ajay Sharma).

Though most of these studies have reported anomalous insertions, abnormal structure or absence of EOMs intra-operatively, with the availability of imaging that can accurately identify EOM, it would seem logical to perform EOM imaging in patients with craniosynostosis and strabismus, prior to planning strabismus surgery. This would help localize the positions of abnormally placed EOMs or document their absence, so that an appropriate surgical plan can be customized to the problem without excessive intra-operative manipulations.

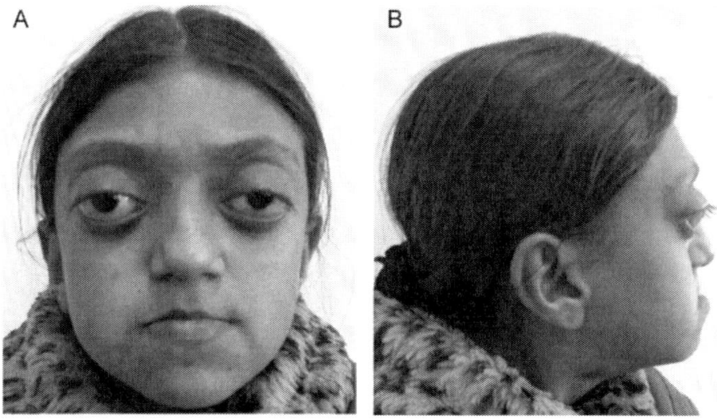

Figure 4. A and B: Typical front and side profile of a patient with Crouzon's syndrome demonstrating ocular proptosis, exotropia, hypertelorism, low-set ears and mid-facial retrusion. (Photo courtesy: Professor Ajay Sharma).

Imaging in Patients with Craniosynostosis and Strabismus

Imaging of the head and orbit is an integral part of the diagnosis and management of patients with craniosynostosis. Multiple useful modalities include simple skiagrams (X-ray skull) that show the shape of the skull along with pressure effects like the classical copper beaten appearance (Figure 1B, 6), computerized tomography (CT) of the skull and orbits augmented by 3D reconstruction (Figure 1 to 3) and magnetic resonance imaging (MRI) of the brain and orbits. High-resolution surface coil MRI has been used for the evaluation of many types of complicated strabismus (Figure 7). [67, 68] It is evident that MRI may be very useful in the evaluation of patients with craniosynostosis and strabismus especially with relation to information about the EOM pulley position and displacement.

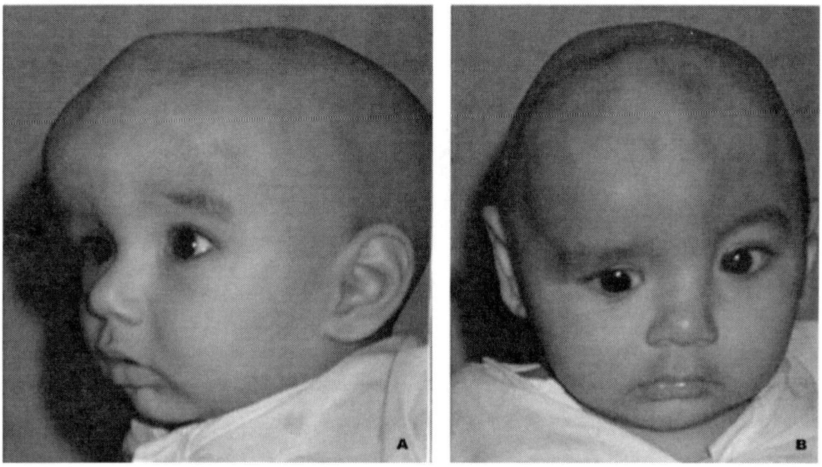

Figure 5. A and B: Side and front profile of a child with left unicoronal craniosynostosis and resultant plagiocephaly who presented with right orbital dystopia and hypotropia. This is an example of anisometropia, aniso-astigmatism and strabismus contralateral to the unicoronal craniosynostosis because of the pressure effects of the growing skull that cannot expand on the synostotic side. (Photo courtesy: Professor Ajay Sharma) .

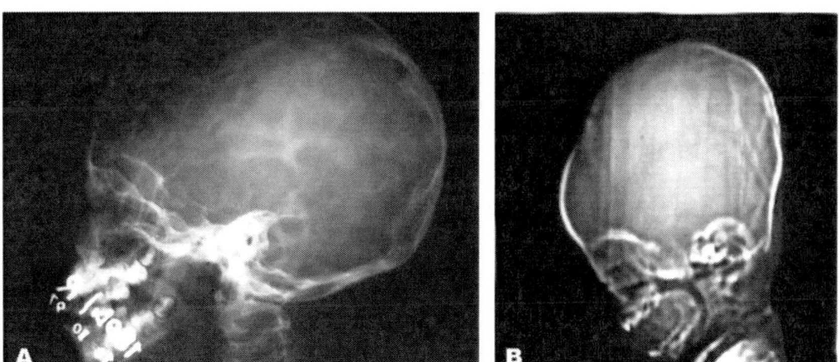

Figure 6. A and B: Simple plain x-rays demonstrate the abnormal skull shapes and the copper beaten appearance of the skull in craniosynostosis. (A) An elongated, boat like skull suggestive of scaphocephaly. (B) Classical "tower skull" appearance in oxycepaly due to multiple suture stenosis, both anteriorly and posteriorly. (Photo courtesy: Professor Ajay Sharma).

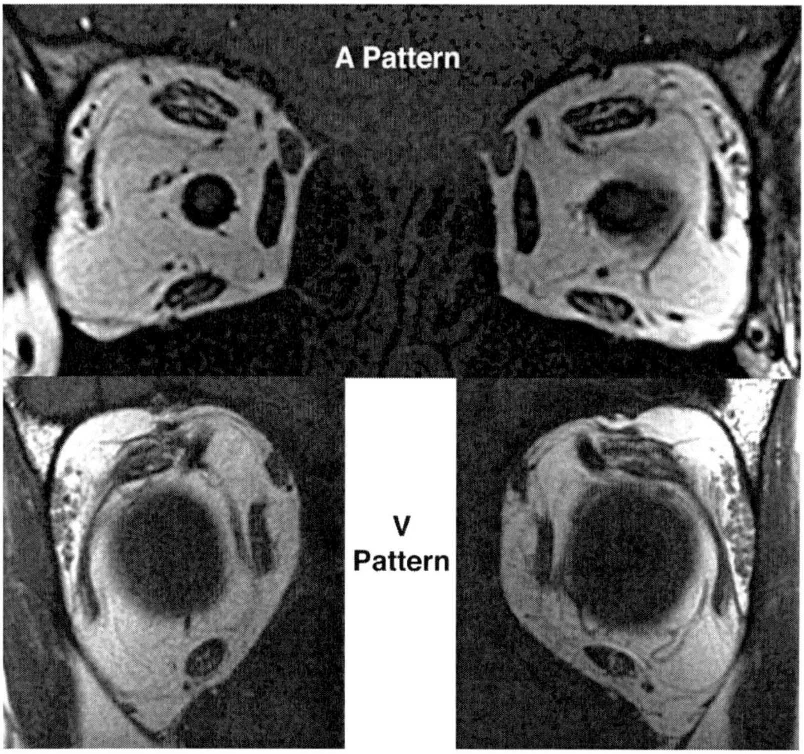

Figure 7. Typical coronal MRI from 2 representative patients showing heterotopic rectus pulleys associated with A-pattern (top) and V-pattern (bottom) strabismus. A-pattern and V-pattern strabismus is commonly observed in patients with craniosynostosis.
Copyright acknowledgement - Elsevier: Demer JL, et al. A 12-Year, Prospective Study of Extraocular Muscle Imaging in Complex Strabismus. J AAPOS 2002; 6 (6): 337-347; Figure 9); Copyright Permission, March 2013.

The high rates of pattern strabismus associated with syndromic craniosynostosis, especially V-pattern with over-elevation in adduction, has suggested etiological hypotheses corroborated by imaging. In one report of a case with craniosynostosis and V-pattern strabismus, CT scan demonstrated that EOMs in one orbit was rotated clock-wise and the other counter-clockwise. [69] This finding was corroborated intra-operatively upon the demonstration of bilateral inferiorly slanted lateral rectus (LR) muscle. [69] In another study, 80% of patients with syndromic craniosynostosis had over-elevation in adduction with a V-pattern strabismus. Examination of the association between the degree of excyclorotation of the rectus EOM array on CT scan or MRI showed that 87.5% patients with over-elevation on adduction had more excyclorotation than age matched controls. [10] Other imaging studies have revealed varying degrees of orbital deformation, excyclorotation, mal-positioning and absence of EOMs that appeared plausibly causal of the type of strabismus present. [5]. A preliminary study using CT scan in unicoronal craniosynostosis with ipsilateral hypertropia revealed superolateral displacement of the superior rectus (SR) muscle pulley. [8] Biomechanical modeling revealed that SR pulley displacement in such cases better accounts for the ipsilateral hypertropia than posterior displacement of the trochlea, with resultant inferior oblique overaction. [8] Knowledge about the exact cause of the strabismus is helpful in formulating appropriate surgical plan.

While CT and MRI have proved useful in determining anomalous position of the EOMs, 3D ultrasonography has been reported useful option for determining the anatomical texture of the affected EOM, including assessment of thin, thick, fibrotic, scarred and absent EOM. Compared to the intra-operative findings, ultrasonography showed a specificity of 88% and sensitivity of 80%. [70]

Management of Strabismus in Craniosynostosis

As the mechanism of strabismus in craniosynostosis is postulated in many cases to be due to the abnormal configuration of the head and orbit, it is plausible that craniofacial surgeries to remodel the skull and face might affect the extent, angle and orientation of strabismus. [3, 18] The goal of craniofacial surgery is aimed to coincide with facial growth and psychosocial development in these children and varies according to the variable expressivity of the condition, although surgical treatment usually begins by 6-12 months of life, sometimes as early as 3 months. [3, 71, 72] Again, multiple operative procedures may be required.

Surgical care in syndromic craniosynostosis involves craniotomy facilitating early release of the stenosed coronal sutures with fronto-orbital advancement to reduce dysmorphic and unwanted skull growth, variants of Le Fort III osteotomy, monobloc frontofacial advancements, bi-parietal release and craniofacial dysjunction. These are often followed by gradual bone distraction to produce correction of associated exophthalmos. [3] Endoscopic strip craniectomy, which can be performed by 3 months of age, has been used as an alternative to the fronto-orbital advancement in children with unicoronal craniosynostosis, with lesser post-operative incidence of V-pattern strabismus and aniso-astigmatism. [18]

Surgical interventions in non-syndromic craniosynostosis are guided by the severity of synostosis and shape of the skull. These comprise craniofacial and calvarial release, and correction of hypertelorism, if present. [3, 72] Management of hypertelorism is an essential part of the functional and cosmetic rehabilitation of the patient, and is also important while treating ocular misalignment. [54]

One study of 11 patients with syndromic craniosynostosis noted spontaneous amelioration of exotropia in 8 patients with Crouzon's syndrome. Esotropia in 2 patients with Apert's syndrome in the same study remained unchanged while vertical deviation somewhat diminished but did not resolve completely following the craniofacial surgery. [73] Recommendations from this study and other similar studies propose performing strabismus surgery at least 3 to 6 months after craniofacial surgery, which should be performed early to avoid complications of visual loss. [14-17, 46, 50, 73] Strabismus re-surgeries may be required, even though strabismus surgery performed in childhood may have been successful, because the strabismus may evolve over time. [22, 47] Strabismus may develop for the first time after craniofacial surgery, sometimes subsequent to nerve paresis; this emphasizes the need to delay strabismus surgeries until the deviation stabilizes after craniofacial surgery. [17, 22, 25, 50, 74, 75] However, other studies have reported infrequent changes in ocular alignment after craniofacial surgery and have recommended that strabismus surgery may be performed prior to craniofacial surgery, in fact as early as possible, for both, attainment of binocularity and also because strabismus surgery after orbital surgery was felt to be technically more difficult and less predictable. [62] There are no clear-cut protocols of standards of practice regarding this. Further study is urgently required.

(A) Non-surgical Management

Standard protocols for correction of refractive errors, astigmatism and amblyopia management apply to patients with craniosynostosis. Appropriate visual, systemic and local examination, assessment of visual acuity for the corresponding age of the patient and anterior and posterior segment assessment is important. [30] Fitting of spectacles may be problematic because of the abnormal facial morphology and is particularly difficult when the child must also wear hearing aids. It is of importance while evaluating strabismus to remember that plagiocephaly may present with facial asymmetry, vertical strabismus on the affected side and head tilt to the sound side, thus simulating a superior oblique (SO) palsy in the affected side but without a paretic muscle. [17, 51]

(B) Surgical Management

Strabismus surgery, as mentioned above, is aided by appropriate pre-operative imaging to determine the cause of the strabismus, especially in the milieu of abnormally located and absent muscles. [8, 29, 57] Computational simulation of ocular motility in craniosynostosis suggests that when orbital excyclorotation occurs in combination with anomalous EOMs, surgical transposition of rectus muscles is usefully combined with surgery on the obliques. [9] Different surgical methods that have been effectively used to correct strabismus in the setting of craniosynostosis include scleral posterior fixation augmenting the Knapp procedure to decrease hypotropia in patients with syndromic craniosynostosis and superior rectus aplasia, [22] anterior and nasal transposition of the inferior oblique (IO) muscle in patients with missing SO muscles, [57] IO denervation and extirpation, and superior suturing of the LR muscle belly to the sclera at a distance of 18 mm from the limbus, converting the LR muscle into an elevator, followed by superior rectus (SR) recession and nasal repositioning to address the V-pattern strabismus. [69] These are special methods described in addition to standard horizontal and vertical EOM surgery, IO recession, IO myectomy, and SO plication, and a combination of rectus and oblique EOM surgery that may be performed. [3, 7, 11, 19, 47] Intra-operative exploration of all six EOMs have been suggested by some authors to evaluate and record frequent anatomic anomalies seen in association with craniosynostosis but, more reasonably, pre-operative orbital imaging would obviate such an invasive recommendation. [58] However, considering the varied causes of ocular mis-alignment in these patients, re-surgeries, both for the craniofacial abnormality and ocular mis-alignment is common and should be explained to the patient before performing surgery.

In conclusion, strabismus in craniosynostosis is difficult to treat, considering the multiple problems that these patients face. Management should be customized to the specific requirement of individual patients. The same clinical presentation of over elevation in adduction may have different structural and functional etiologies in the form of absent SO muscle, anomalous SO muscle, dysfunctional or paretic SO muscle, infra-placed LR muscle, posteriorly located SR pulley or simply, secondary IO over action associated with poor vision in the affected eye. EOM imaging is a valuable pre-operative diagnostic test in such a milieu. A multi-disciplinary team comprising neuro-surgeons, plastic surgeons, maxilla-facial surgeons, ophthalmologists, pediatricians, and otolaryngologists is valuable in providing appropriate support to ensure optimum rehabilitation.

References

[1] Wolfgang D. Differential diagnosis of skull and spine disorders. In: John J, ed. *Radiology Review Manual*: Lippincott, Wilkins and Williams; 2002:174.

[2] Boyadjiev SA. Genetic analysis of non-syndromic craniosynostosis. *Orthod Craniofac Res.* 2007;10(3):129-137.

[3] Rosenberg JB, Tepper OM, Medow NB. Strabismus in Craniosynostosis. *J Pediatr Ophthalmol Strabismus.* 2012:1-9.

[4] Dufier JL, Vinurel MC, Renier D, Marchac D. [Ophthalmologic complications of craniofacial stenoses. Apropos of 244 cases]. *J Fr Ophtalmol.* 1986;9(4):273-280.

[5] Cheng H, Burdon MA, Shun-Shin GA, Czypionka S. Dissociated eye movements in craniosynostosis: a hypothesis revived. *Br J Ophthalmol.* 1993;77(9):563-568.

[6] Macintosh C, Wall S, Leach C. Strabismus in unicoronal synostosis: ipsilateral or contralateral? *J Craniofac Surg.* 2007;18(3):465-469.

[7] Holmes JM, Hatt SR, Leske DA. Superior oblique tucks for apparent inferior oblique overaction and V-pattern strabismus associated with craniosynostosis. *Strabismus.* 2010;18(3):111-115.

[8] Weiss AH, Phillips JO. Hypertropia associated with superolateral translation of the superior rectus muscle pulley in unilateral coronal synostosis. *Arch Ophthalmol.* 2006;124(8):1128-1134.
[9] Clement R, Nischal K. Simulation of oculomotility in Craniosynostosis patients. *Strabismus.* 2003;11(4):239-242.
[10] Tan KP, Sargent MA, Poskitt KJ, Lyons CJ. Ocular overelevation in adduction in craniosynostosis: is it the result of excyclorotation of the extraocular muscles? *J Aapos.* 2005;9(6):550-557.
[11] Lehman S. Strabismus in craniosynostosis. *Curr Opin Ophthalmol.* 2006;17(5):432-434.
[12] Newman SA. Ophthalmic features of craniosynostosis. *Neurosurg Clin N Am.* 1991;2(3):587-610.
[13] Hertle RW, Quinn GE, Minguini N, Katowitz JA. Visual loss in patients with craniofacial synostosis. *J Pediatr Ophthalmol Strabismus.* 1991;28(6):344-349.
[14] Denis D, Genitori L, Conrath J, Lena G, Choux M. Ocular findings in children operated on for plagiocephaly and trigonocephaly. *Childs Nerv Syst.* 1996;12(11):683-689.
[15] Denis D, Genitori L, Bolufer A, Lena G, Saracco JB, Choux M. Refractive error and ocular motility in plagiocephaly. *Childs Nerv Syst.* 1994;10(4):210-216.
[16] Denis D, Genitori L, Bardot J, Saracco JB, Choux M, Maumenee IH. Ocular findings in trigonocephaly. *Graefes Arch Clin Exp Ophthalmol.* 1994;232(12):728-733.
[17] Gosain AK, Steele MA, McCarthy JG, Thorne CH. A prospective study of the relationship between strabismus and head posture in patients with frontal plagiocephaly. *Plast Reconstr Surg.* 1996;97(5):881-891.
[18] MacKinnon S, Rogers GF, Gregas M, Proctor MR, Mulliken JB, Dagi LR. Treatment of unilateral coronal synostosis by endoscopic strip craniectomy or fronto-orbital advancement: Ophthalmologic findings. *J Aapos.* 2009;13(2):155-160.
[19] Coats DK, Ou R. Anomalous medial rectus muscle insertion in a child with craniosynostosis. *Binocul Vis Strabismus Q.* 2001;16(2):119-120.
[20] Coats DK, Paysse EA, Stager DR. Surgical management of V-pattern strabismus and oblique dysfunction in craniofacial dysostosis. *J Aapos.* 2000;4(6):338-342.
[21] Carruthers JD. Strabismus in craniofacial dysostosis. *Graefes Arch Clin Exp Ophthalmol.* 1988;226(3):230-234.

[22] Rattigan S, Nischal KK. Foster-type modification of the Knapp procedure for anomalous superior rectus muscles in syndromic craniosynostoses. *J Aapos.* 2003;7(4):279-282.

[23] Elmslie FV, Reardon W. Craniofacial developmental abnormalities. *Curr Opin Neurol.* 1998;11(2):103-108.

[24] Tandon A. Craniofacial Malformations. In: Chaudhuri Z, Vanathi M, eds. *Postgraduate Ophthalmology.* New Delhi: Jaypee Highlights Medical Publishers, Inc; 2012:1864-1873.

[25] Anderson FM. Treatment of coronal and metopic synostosis: 107 cases. *Neurosurgery.* 1981;8(2):143-149.

[26] Bruner E. Cranial shape and size variation in human evolution: structural and functional perspectives. *Childs Nerv Syst.* 2007;23(12):1357-1365.

[27] Lyons CJ. Craniofacial Abnormalities. In: Taylor D, Hyot C, eds. *Pediatric Ophthalmology and Strabismus.* Vol 3. Edinburgh: Elsevier Ltd; 2005:354-367.

[28] Vidal Sanahuja R, Gean Molins E, Sanchez Garre C, Quilis Esquerra J, Garcia Fructuoso G, Costa Clara JM. [Presentation of two cases of Crouzon syndrome: allelic cranio-stenotic conditions of FGFR genes]. *An Pediatr (Barc).* 2012;77(4):272-278.

[29] Kerr NC. Management of vertical deviations secondary to other anatomical abnormalities. *Am Orthopt J.* 2011;61:39-48.

[30] Baranello G, Vasco G, Ricci D, Mercuri E. Visual function in nonsyndromic craniosynostosis: past, present, and future. *Childs Nerv Syst.* 2007;23(12):1461-1465.

[31] Tarczy-Hornoch K, Smith B, Urata M. Amblyogenic anisometropia in the contralateral eye in unicoronal craniosynostosis. *J Aapos.* 2008;12(5):471-476.

[32] Jadico SK, Huebner A, McDonald-McGinn DM, Zackai EH, Young TL. Ocular phenotype correlations in patients with TWIST versus FGFR3 genetic mutations. *J Aapos.* 2006;10(5):435-444.

[33] Jadico SK, Young DA, Huebner A, et al. Ocular abnormalities in Apert syndrome: genotype/phenotype correlations with fibroblast growth factor receptor type 2 mutations. *J Aapos.* 2006;10(6):521-527.

[34] Khong JJ, Anderson PJ, Hammerton M, Roscioli T, Selva D, David DJ. Differential effects of FGFR2 mutation in ophthalmic findings in Apert syndrome. *J Craniofac Surg.* 2007;18(1):39-42.

[35] Levy RL, Rogers GF, Mulliken JB, Proctor MR, Dagi LR. Astigmatism in unilateral coronal synostosis: incidence and laterality. *J Aapos.* 2007;11(4):367-372.

[36] Tay T, Martin F, Rowe N, et al. Visual manifestations of craniofrontonasal dysplasia. *J Pediatr Ophthalmol Strabismus.* 2007;44(4):251-254.

[37] Khong JJ, Anderson P, Gray TL, Hammerton M, Selva D, David D. Ophthalmic findings in apert syndrome prior to craniofacial surgery. *Am J Ophthalmol.* 2006;142(2):328-330.

[38] Khong JJ, Anderson P, Gray TL, Hammerton M, Selva D, David D. Ophthalmic findings in Apert's syndrome after craniofacial surgery: twenty-nine years' experience. *Ophthalmology.* 2006;113(2):347-352.

[39] Tay T, Martin F, Rowe N, et al. Prevalence and causes of visual impairment in craniosynostotic syndromes. *Clin Experiment Ophthalmol.* 2006;34(5):434-440.

[40] Oliveira NA, Alonso LG, Fanganiello RD, Passos-Bueno MR. Further evidence of association between mutations in FGFR2 and syndromic craniosynostosis with sacrococcygeal eversion. *Birth Defects Res A Clin Mol Teratol.* 2006;76(8):629-633.

[41] Gupta PC, Foster J, Crowe S, Papay FA, Luciano M, Traboulsi EI. Ophthalmologic findings in patients with nonsyndromic plagiocephaly. *J Craniofac Surg.* 2003;14(4):529-532.

[42] Greenberg MF, Pollard ZF. Ocular plagiocephaly: ocular torticollis with skull and facial asymmetry. *Ophthalmology.* 2000;107(1):173-178; discussion 178-179.

[43] Cohen MM, Jr., Kreiborg S. A clinical study of the craniofacial features in Apert syndrome. *Int J Oral Maxillofac Surg.* 1996;25(1):45-53.

[44] Freihofer HP. [Syndromes 6. Crouzon's disease]. *Ned Tijdschr Tandheelkd.* 1998;105(12):451-452.

[45] Fredrick DR, Mulliken JB, Robb RM. Ocular manifestations of deformational frontal plagiocephaly. *J Pediatr Ophthalmol Strabismus.* 1993;30(2):92-95.

[46] Buncic JR. Ocular aspects of Apert syndrome. *Clin Plast Surg.* 1991;18(2):315-319.

[47] Hsu CM, Lin MC, Sheu SJ. Manifested strabismus in a case of Apert syndrome. *J Chin Med Assoc.* 2011;74(2):95-97.

[48] Khan SH, Nischal KK, Dean F, Hayward RD, Walker J. Visual outcomes and amblyogenic risk factors in craniosynostotic syndromes: a review of 141 cases. *Br J Ophthalmol.* 2003;87(8):999-1003.

[49] Inagaki T, Kyutoku S, Seno T, et al. The intracranial pressure of the patients with mild form of craniosynostosis. *Childs Nerv Syst.* 2007;23(12):1455-1459.
[50] Muhling J, Zoller J, Saffar M, Reinhart E, Reuther J. [Results of surgical therapy of orbital abnormalities]. *Mund Kiefer Gesichtschir.* 1998;2 Suppl 1:S94-97.
[51] Hromadkova L, Rehurek J, Uherkova E. [Vertical strabismus in plagiocephaly]. *Cesk Oftalmol.* 1992;48(3):181-185.
[52] Mohan H, Sen DK, Gupta DK. Scaphocephaly, bilateral aniridia, and bilateral anterior polar cataract with posterior dislocation of the lens. *Br J Ophthalmol.* 1971;55(9):642-643.
[53] Pillay VK. Acrocephalosyndactyly in Singapore. A Study of Five Chinese Males. *J Bone Joint Surg Br.* 1964;46:94-101.
[54] Cohen MM, Jr., Richieri-Costa A, Guion-Almeida ML, Saavedra D. Hypertelorism: interorbital growth, measurements, and pathogenetic considerations. *Int J Oral Maxillofac Surg.* 1995;24(6):387-395.
[55] Macintosh C, Wells R, Johnson D, Wall S. What are the effects of metopic synostosis on visual function? *J Craniofac Surg.* 2011;22(4):1280-1283.
[56] Dickmann A, Parrilla R, Aliberti S, et al. Prevalence of neurological involvement and malformative/systemic syndromes in A- and V-pattern strabismus. *Ophthalmic Epidemiol.* 2012;19(5):302-305.
[57] Hussein MA, Stager DR, Sr., Beauchamp GR, Stager DR, Jr., Felius J. Anterior and nasal transposition of the inferior oblique muscles in patients with missing superior oblique tendons. *J Aapos.* 2007;11(1):29-33.
[58] Greenberg MF, Pollard ZF. Absence of multiple extraocular muscles in craniosynostosis. *J Aapos.* 1998;2(5):307-309.
[59] Tomac S, Kalayci D, Hasiripi H. Synostotic plagiocephaly causing pseudoparalysis of the superior oblique and ocular torticollis: report of a case with unique sensory findings. *Binocul Vis Strabismus Q.* 1998;13(4):267-272.
[60] Bustos DE, Donahue SP. Absence of all cyclovertical extraocular muscles in a child who has Apert syndrome. *J Aapos.* 2007;11(4):408-409.
[61] Cuttone JM, Brazis PT, Miller MT, Folk ER. Absence of the superior rectus muscle in Apert's syndrome. *J Pediatr Ophthalmol Strabismus.* 1979;16(6):349-354.

[62] Diamond GR, Katowitz JA, Whitaker LA, Quinn GE, Schaffer DB. Variations in extraocular muscle number and structure in craniofacial dysostosis. *Am J Ophthalmol.* 1980;90(3):416-418.
[63] Margolis S, Pachter BR, Breinin GM. Structural alterations of extraocular muscle associated with Apert's syndrome. *Br J Ophthalmol.* 1977;61(11):683-689.
[64] Snir M, Gilad E, Ben-Sira I. An unusual extraocular muscle anomaly in a patient with Crouzon's disease. *Br J Ophthalmol.* 1982;66(4):253-257.
[65] Pollard ZF. Bilateral superior oblique muscle palsy associated with Apert's syndrome. *Am J Ophthalmol.* 1988;106(3):337-340.
[66] Sharma P, Prakash P. Degsagittalisation of obliques: a case report of plagiocephaly with inferior oblique overaction. *Indian J Ophthalmol.* 1993;41(1):32-34.
[67] Shin GS, Demer JL, Rosenbaum AL. High resolution, dynamic, magnetic resonance imaging in complicated strabismus. *J Pediatr Ophthalmol Strabismus.* 1996;33(6):282-290.
[68] Demer JL, Clark RA, Kono R, Wright W, Velez F, Rosenbaum AL. A 12-year, prospective study of extraocular muscle imaging in complex strabismus. *J Aapos.* 2002;6(6):337-347.
[69] Velez FG, Thacker N, Britt MT, Rosenbaum AL. Cause of V pattern strabismus in craniosynostosis: a case report. *Br J Ophthalmol.* 2004;88(12):1598-1599.
[70] Somani S, Mackeen LD, Morad Y, et al. Assessment of extraocular muscles position and anatomy by 3-dimensional ultrasonography: a trial in craniosynostosis patients. *J Aapos.* 2003;7(1):54-59.
[71] Wong GB, Kakulis EG, Mulliken JB. Analysis of fronto-orbital advancement for Apert, Crouzon, Pfeiffer, and Saethre-Chotzen syndromes. *Plast Reconstr Surg.* 2000;105(7):2314-2323.
[72] Whitaker LA, Bartlett SP. The craniofacial dysostoses: guidelines for management of the symmetric and asymmetric deformities. *Clin Plast Surg.* 1987;14(1):73-81.
[73] Morax S. Change in eye position after cranio-facial surgery. *J Maxillofac Surg.* 1984;12(2):47-55.
[74] Anderson PJ, David DJ. Late results after unicoronal craniosynostosis correction. *J Craniofac Surg.* 2005;16(1):37-44.
[75] Anderson PJ, Tan E, David DJ. Simultaneous multiple vector distraction for craniosynostosis syndromes. *Br J Plast Surg.* 2005;58(5):626-631.

In: Craniofacial Disorders
Editors: Sh.B. Turner and J. E. Miller
ISBN: 978-1-62618-923-2

Chapter 5

Cri du Chat Syndrome: Craniofacial and Orofacial Features and Treatment Considerations

R. M. Yáñez-Vico, D. Torres-Lagares, A. Rodriguez-Caballero and G. Machuca-Portillo
Department of Stomatology.University of Seville, Spain

Abstract

Cri du Chat syndrome (CdCS) is a genetic disease resulting from a partial or total deletion on the short arm of chromosome 5 (5p-). CdCS is one of the most frequent chromosomal syndromes in humans. The prevalence has been estimated at 1 in 37.000 live-born infants.

The most characteristic and clinical feature in newborns is a high-pitched monochromatic cat-like crying, that was attributed to anatomical alterations in the larynx and epiglottis. Additional clinical features are severe psychomotor and mental retardation, growth delay, and hypotonia. CdCS causes craniofacial abnormalities such as microcephaly, slight hypertelorism, epicanthal folds, downslanting palpebral fissures, strabismus, malformed ears, mandibular microretrognathism, anterior open-bite and high palate. Buccodental pathology and dental anomalies of patients with this syndrome reported are hypodontia, enamel hypoplasies, dental opacities, tooth agenesis, supernumerary tooth, root resortion, macrodontia, and transposed teeth.

Clinical dentistry management is linked to behavioral patterns. In general, CdC patients have euphoric, cheerful and sociable personalities. However, social isolation and reserved and emotionally detached character have been found in some cases of patients. Because of this and influenced for dental pathology and treatment, dental treatment is suggested to be ambulatory (without deep sedation) or, in contrast, with general anesthesia.

Introduction

Cri du Chat syndrome (CdCS) was first described in 1963 because a high-pitched crying similar to the mewing of a cat was the most characteristic clinical feature [1]. The syndrome is a rare genetic autosomal disorder resulting from a partial or total deletion on the short arm of chromosome 5 (5p-) [2,3]. The size of the deletion could affect from region 5p 15.3 to the complete loss of the short arm. Prevalence of CdCs is estimated at 1 in 37.000 live-born infants, being slightly more prevalent among females (3:1) [3]. Furthermore, Schinzel's data [4] suggest that it is the most common autosomal deletion syndrome in humans.

As it has been mentioned previously, the most characteristic clinical feature is the weak high-pitched crying during the childhood. However, this cry is not pathognomonic of the syndrome because the cat-like cry without the syndrome's typical dysmorphic and severe developmental features has been found in individuals whose deletion is limited to 5p15.3 [3]. The pathogenesis of this cry has been associated with anatomical alterations in the epiglottis (flappy, small and curved) and the larynx (hypoplastic, narrow and diamond shaped or quadrangular) and with structural and/or functional neurological alterations [3]. Additional clinical features are severe psychomotor and severe mental retardation, growth delay and hypotonia [5]. CdCS newborns are usually low birth weight and small size and because modified suction and hypotonia, gastroesophageal reflux difficult the correct feeding and maintained the growth and developmental retardation [3]. Other clinical features include auditive sensitive alterations like hypersensibility to sounds and hypoacusia to high frequencies. Strabismus, myopia, cataracta and hypersensitivity of the pupil to metacholine and resistence to mydriatics are also common anomalies in this syndrome [6]. Other possible alterations less frequent include cardiac, skeletal, genitourinary, metabolic and immune defects [1,3,7].

Craniofacial Features

Several authors have been described clinical features that include microcephaly, slight hypertelorism, epicanthal folds, down-slanting palpebral fissures, strabismus and malformed ears [8,9,10]. During the first years of life, moon face appears about 80% of CdCS children but, because vertical growth pattern, the face turn into more long than width face [8]. Down-turned corners of the mouth (81%), large nasal bridge (87.2%), short philtrum (87.8%), convex facial profile with mandibular microretrognathia (96.7%), abnormal dermatoglyphics (92%) and premature grey hair are other alterations widely described [1,3,8,10] (*figure 1*).

In our clinical observational study on 33 CdCS patients we found epicanthus and hyperthelorism in most of subjects (93.75%). Some authors consider these features as subjective, because of a visual effect the flat wide nasal bridge provokes (93.75%). According to Nieburh, it would be more precise to talk about "eyes slightly separated" [3], and he agrees with Sedano and Howard when he states that true hypertelorism is a rare characteristic of cri-du-chat syndrome [6,11].We also found high prevalence of other common craniofacial features described previously as short philtrum, low-set ears, strabism and convex profile [10]. These data come from the observation of the patient and the measurement in facial photographs.

Figure 1. Typical facial characteristics in CdCS: slight hypertelorism, epicanthal folds, down-slanting palpebral fissures, strabismus, down-turned corners of the mouth and large nasal bridge.

Hovewer, we also conducted other study in which lateral head profile radiographs were measurement [12]. To evaluate and quantify relationships between the jaws, between the jaws and the cranial bones and between soft tissues with bones and teeth we use cephalometry. Vertical facial pattern were found in the decreases upper gonial angle and increased lower gonial angle and a larger distance between the Nasion and Menton landmarks (increased anterior facial height) (*Figure 2*). However, other parameters that evaluate facial pattern showed mesofacialcharacteristics. One of the most characteristic features in analysed subjects was the inclined palatal plane, with 70% of the subjects with a marked slope. Data previously mentioned in the available literature emphasised decreased long bone development. This agrees with our findings since our data reveal decreased length in the anterior and posterior cranial base. Furthermore, total maxillary length and mandibular corpus are decreased. Skeletal class II malocclusion is a common feature in CdC patients, as it results from a protruding maxilla that worsens the condition because of a short anterior cranial base and mandibular retrognathism influenced by a larger cranial base angle [12]. Other characteristic feature is dental biprotusion. This means that upper and lower incisor are toward inclined causing the typic facial convex profile and protruded lips.

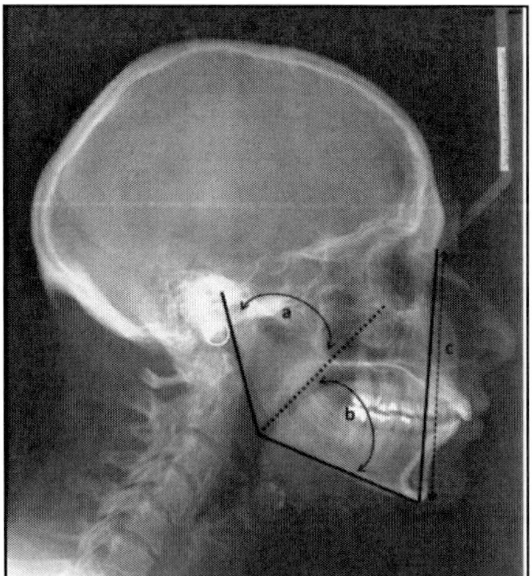

Figure 2. Lateral head profile radiography of a CdCS patient. a) uppergonial angle; b) lower gonial angle; c) anterior facial height (distance between Nasion and Menton landmarks).

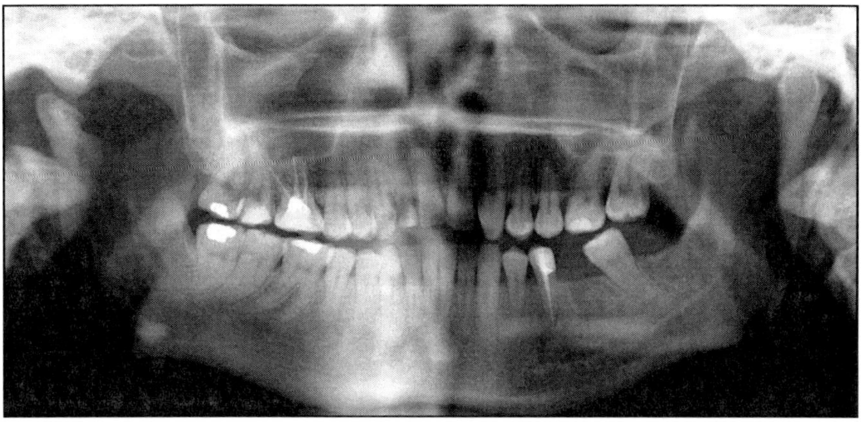

Figure 3. Ortopantomograph of a CdCS patient. Agenesis of left lateral superior incisor, generalized mild root resorptionare shown.

Oral Features

There are no many data about oral features in CdCS. The main oral and dental anomalies registered were enamel hypoplasia [11], poor oral hygiene, generalized chronic periodontitis and retardation of tooth eruption [13,14]. In spite of these alterations, there is not enough evidence to support any clear association between these pathologies and the CdCS. In our clinical observational study on 33 CdCS patients, we found dental anomalies (56.72%) such as enamel hypoplasies(54%), dental opacities (33%), tooth agenesis (9%), supernumerary tooth (3%), root resortion(3%), macrodontia(6%), and transposed teeth (3%)[10] (*Figure 3*).

Because dysphagia (51.61%) and gastroesophageal regurgitation (65.62%) were frequently found in these patients, excessive dental erosions provoked by gastroesophageal reflux and attritions owing to intense day-and-night bruxism were also frequently found [10]. Healthy periodontal condition was rarely found (22.9%) and more than half of them showed tartar in one or more quadrants (52.69%). We also found 4- to 5-mm pouches in young subjects (14.58%) and pouches larger than 5 mm in three of the patients being studied (4.16%)[10].

Prognosis and Dental Treatment Considerations

In spite of 75% of the registered deaths occurred during the first months after birth and 90% during the first year of life [2,3], survival rate and life expectation are relatively high after first years of life [8]. The type of the deletion and the size and location of the deletion determine a typical CdCS phenotype or, in contrast, an atypical clinical profile, influencing the life prognosis of the patient [3,8]. The individuals with atypical CdCS do not present a frequent severe handicap in learning [15], instead they usually present milder anomalies associated with a better prognosis. However, the individuals with typical CdCS usually present most of the characteristics associated with the syndrome and a less hopeful prognosis [8].

Clinical dentistry management is linked to behavioral patterns. Despite the scarcity of studies about the language skills, communication abilities and social interaction, most of the reported cases indicate that the individuals with CdCS generally have a nice and loving personality[8, 15]. In general, CdC patients have euphoric, cheerful and sociable personalities. However, social isolation and reserved and emotionally detached character have been found in some cases of patients. Hyperactivity, loss of attention, uneasiness, aggressive and self-injured behavior are been described as the principal behavioral problems [8, 16]. We found high incidence (77.8%) of euphoric, cheerful, and sociable personalities among the 33 patients evaluated, whereas only a reduced number of them had a reserved and emotionally detached character[10]. Because of this and influenced for dental pathology and treatment, dental treatment is suggested to be ambulatory (without deep sedation) or, in contrast, with general anesthesia. On the other hand, it is important to consider the high prevalence of self-injured behavior (hitting their head against objects, hitting the head against body parts, and self-injuries caused by biting) [10,15]. This determined in our CdCS population the treatment under deep sedation or general anesthesia, and only a small number of them could be treated using movement restriction techniques[10]. The first appointment in the dental clinic determine treatment necessities according to the classification of the conduct according to the FranklScale [17], and Scale for Clinical Management of the Patient [18]. The severity of dental pathologies is other important point to consider. Complex treatments that need several and large appointments could be with general anaesthesia in only one. In contrast, if the patient needs minimum or preventive treatment could be not necessary general anaesthesia and restriction techniques would be useful.

References

[1] Lejeune, J; Lafourcade, J; Berger, R; Vialatte, J; Boeswillwald, M; Seringe, P; et al. 3 cases of partial deletion of the short arm of a 5 chromosome. *C R Hebd Seances Acad Sci.*, 1963, 257, 3098-102.

[2] Mainardi, PC; Perfumo, C; Calì, A; Coucourde, G; Pastore, G; Cavani, S; et al. Clinical and molecular characterisation of 80 patients with 5p deletion: genotype-phenotype correlation. *J Med Genet.*, 2001, 38, 151-8.

[3] Niebuhr, E. The Cri du Chat syndrome: epidemiology, cytogenetics, and clinical features. *Hum Genet.*, 1978, 44, 227-75.

[4] Schinzel, A. Catalogue of unbalanced chromosome aberrations in man, 2nd edition. Berlin, Germany. Walter de Gruyter& Co: 1984.

[5] Overhauser, J; Huang, X; Gersh, M; Wilson, W; McMahon, J; Bengtsson, U; et al. Molecular and phenotypic mapping of the short arm of chromosome 5, sublocalization of the critical region for the cri-du-chat syndrome. *Hum Mol Genet*, 1994, 3, 247-52.

[6] Howard, RO. Ocular abnormalities in the cri du chat syndrome. *Am J Ophthalmol.*, 1972, 73, 949-54.

[7] Takebayashi, T; Obata, H; Minaki, Y; Sekine, M; Imoto, K; Yokogushi, K; et al. Scoliosis in catcrysyndrome. *J Orthop Sci.*, 2006, 11, 259- 63.

[8] Rodríguez-Caballero, A; Torres-Lagares, D; Rodríguez-Pérez, A; Serrera-Figallo, MA; Hernández-Guisado, JM; Machuca-Portillo, G. Cri du chat syndrome: a criticalreview. *Med Oral Patol Oral Cir Bucal.*, 2010, 15, e473-8.

[9] Mainardi, PC. Cri du Chat syndrome. *Orphanet J Rare Dis*, 2006, 1, 33-42.

[10] Rodríguez-Caballero, A; Torres-Lagares, D; Yáñez-Vico, RM; Gutiérrez-Pérez, JL; Machuca-Portillo, G. Assessment of orofacial characteristics and oral pathology associated with cri-du-chat syndrome. *Oral Dis.*, 2012, 18, 191-7.

[11] Sedano, HO; Look, RA; Carter, C; Cohen, MM; Jr. B group short-arm deletion syndrome. *Birth Defects OrigArtic Ser.*, 1971, 7, 89-97.

[12] Yáñez-Vico, RM; Rodríguez-Caballero, A; Iglesias-Linares, A; Guerra-López, N; Torres-Lagares, D; Machuca-Portillo, G; Solano-Reina, E;

Gutiérrez-Pérez, JL. Craniofacial characteristics in cri-du-chat syndrome. *Oral Surg Oral Med Oral Pathol Oral Radiol Endod.*, 2010, 110, e38-44.

[13] Scully, C; Davison, MF. Orofacial manifestations in the Cri du Chat syndrome (5p-). *J Dent.*, 1979, 7, 313-20.

[14] Boraz, RA. Cri-du-chat syndrome: dental considerations and report of case. *Spec Care Dentist.*, 1990, 10, 13-5.

[15] Cornish, KM; Bramble, D; Munir, F; Pigram, J. Cognitive functioning in children with typical cri du chat (5p-) syndrome. *Dev Med Child Neurol.*, 1999, 41, 263-6.

[16] Collins, MS; Cornish, K. A survey of the prevalence of stereotypy, self-injury and aggression in children and young adults with Cri du Chat syndrome. *J Intellect Disabil Res.*, 2002, 46, 133-40.

[17] Frankl, SN; Shiere, FR; Fogelo, HR. Should remain with the child in the dental operatory? *Dent Child*, 1962, 29, 150–163.

[18] Bensberg, GJ; Barnett, CD; Menius, JA. A survey: dental services in state residential facilities. *Ment Retard*, 1966, 4, 8–13.

Index

A

B

C

F

G

H

I

J

K

L

M

N

O

P

Q

R

S